NUCLEAR WEAPONS

T0332151

The MIT Press Essential Knowledge series

A complete list of the titles in this series appears at the back of this book.

NUCLEAR WEAPONS

MARK WOLVERTON

The MIT Press | Cambridge, Massachusetts | London, England

The MIT Press would like to thank the anonymous peer reviewers who provided comments on drafts of this book. The generous work of academic experts is essential for establishing the authority and quality of our publications. We acknowledge with gratitude the contributions of these otherwise uncredited readers.

This book was set in Chaparral Pro by New Best-set Typesetters Ltd. Printed and bound in the United States of America.

Library of Congress Cataloging-in-Publication Data

Names: Wolverton, Mark, author.
Title: Nuclear weapons / Mark Wolverton.
Description: Cambridge, Massachusetts : The MIT Press, [2021] | Series: The MIT Press essential knowledge series | Includes bibliographical references and index.
Identifiers: LCCN 2021010580 | ISBN 9780262543316 (paperback)
Subjects: LCSH: Nuclear weapons—History. | Deterrence (Strategy)
Classification: LCC U264 .W65 2021 | DDC 355.8/2511909—dc23
LC record available at https://lccn.loc.gov/2021010580

10   9   8   7   6   5   4   3   2   1

CONTENTS

SERIES FOREWORD

The MIT Press Essential Knowledge series offers accessible, concise, beautifully produced pocket-size books on topics of current interest. Written by leading thinkers, the books in this series deliver expert overviews of subjects that range from the cultural and the historical to the scientific and the technical.

In today's era of instant information gratification, we have ready access to opinions, rationalizations, and superficial descriptions. Much harder to come by is the foundational knowledge that informs a principled understanding of the world. Essential Knowledge books fill that need. Synthesizing specialized subject matter for nonspecialists and engaging critical topics through fundamentals, each of these compact volumes offers readers a point of access to complex ideas.

PREFACE

Since their birth into a war-ravaged world in 1945, nuclear weapons have been part of our world, fears, politics, and the fabric of our culture. At first they were exotic, mysterious, and terrifying, and then a constant threat throughout the many years of the Cold War. But now in the twenty-first century, they've become almost mundane, boring, and irrelevant—something most folks rarely even think about, much less worry about or fear.

Yet they continue to exist, they still pose an existential threat, and even if the prospect of ultimate doom they once represented has greatly receded (though hardly disappeared), in some very real ways they pose greater dangers—and a greater likelihood of being used—than during the Cold War. And with all that, they remain as mysterious and misunderstood as when they were brand new, and perhaps even more shrouded in misconceptions and ignorance as to their true nature and capabilities, with a public whose awareness and knowledge derives more from movies and TV shows than from actual fact.

This book is an attempt to remedy that unfortunate disconnect between common perceptions and reality, between the Bomb as a cultural icon and the hard truth of the thousands of nuclear weapons that still, at this moment, rest in missile silos, storage bunkers, submarines cruising the oceans, the wings and bomb bays of airplanes, and nameless secret places, ready and patiently waiting for the moment they will be called to terrible action.

If nuclear weapons are ever again used, they may not directly affect more than a relative handful of people (though a handful measured in the thousands), but the indirect effects will shake the foundations of every facet of human society. That's why it's important for everyone to have a basic understanding of nuclear weapons: what they are, how and why they originated, the history of attempts to control, limit, and possibly use them, and the prospects for eliminating them completely or at least learning to live with them.

The subject of nuclear weapons is vast and complex, encompassing not just the obvious disciplines of physics and military strategy but also including politics, sociology, psychology, and nearly every form of popular culture. It's impossible to do it all justice in a slim volume such as this one, and in the pages that follow, I won't even try. Instead, my intention is to give the reader a general overview of the subject, including an understanding of the basic scientific facts, some historical perspective, and an appreciation of the unique political, social, and moral dilemmas that surround it all. I also hope to provide a beginning framework for those who wish to pursue individual aspects of this broad subject in greater depth by pointing the way to further resources to consider and explore.

Talk of things such as deterrence theory, fission-fusion warheads, counterforce versus countervalue targeting, and isotope separation can seem hopelessly technical as well as esoteric to the average person who's simply trying to live a happy, peaceful, and productive life without thinking about the prospect of some terrible nuclear doomsday. It's an understandable feeling, and I certainly don't recommend to anyone a full-time mental diet of obsessing about the end of the world.

Yet the unfortunate truth is that far from being a thing of the past, remnant of history, or lurid fictional plot device, nuclear weapons remain an issue that involves every human being. That includes not only every citizen of the world's current nuclear powers but those in the most remote corners of the globe too, from the poles to the tiniest Pacific island. Everyone born since the dawn of the nuclear age has already been physically affected, however subtly and indirectly, by nuclear weapons, as the isotopic by-products of 520 open-air bomb tests have spread over the entire planet to settle in the earth's biosphere, whether deep in our bone marrow, our plants and water, or even our table wines. And the presence of nuclear weapons in the world has profoundly influenced our political and economic structures, our military strategies and tactics, the ways in which nations interact, cooperate, and conflict with each other, and our individual and collective culture and consciousness, all the way down to our everyday vocabularies, imagery, and the nightmares that haunt our sleep.

This book is an attempt to provide a primer for understanding a world with nuclear weapons so that you can better understand the next politician arguing to abandon a treaty or adopt a new weapons system; distinguish legitimate dangers from political rhetoric, scaremongering, and saber-rattling, and appreciate the true meaning of flippant remarks about "nuking" an enemy. Most of us don't like to spend much, if any, time thinking about nuclear warfare or terrorism. But sometimes it's necessary to do so, and at such times, a clear understanding of the subject is important, not only so we can ensure that the leaders who speak and act for us are preserving our safety instead of leading us to disaster, but to calm our own dark, human fears. Because while it may not be our fingers on the nuclear buttons, we all have a profound stake in what happens if those buttons are ever pushed.

This project would never have happened were it not for the enthusiastic support of my editor at the MIT Press, Jermey Matthews. Haley Biermann of the MIT Press patiently endured my endless questions. And of course, my brilliant agent, Michelle Tessler, brought it all together. Many thanks to you all.

INTRODUCTION:
THE BIG DIFFERENCE

In the twenty-first century, we're so used to the idea of nuclear technology that it's become part of our everyday discourse. We talk about "nuking" a slice of pizza. We joke about people and things "glowing in the dark" from living near a nuclear power plant, or mutated three-eyed fish from radioactive waste dumped in the river.

Such casual familiarity extends to nuclear weapons. When the United States gets into a political scuffle or even minor military conflict with another nation, some less-refined types can usually be heard joking that "maybe we should just nuke 'em." As I'll examine later, atomic bombs, hydrogen bombs, intercontinental missiles, terrorist devices, and similar nuclear threats serve as plot devices in endless movies, TV shows, comics, and other popular culture products. They can be deadly serious heralds of doomsday, opportunities for humorous exaggeration, or something in between, but everyone knows the basics. Big explosions. Radiation and fallout. Horrible sickness and mutations. Bad stuff.

And yet aside from the radiation and fallout aspect, most people have little idea of exactly what a nuclear weapon *is*. What makes it so different from other kinds of bombs? Aren't they just bigger and noisier (and maybe somewhat dirtier and nastier because of that radiation thing)? Why are they such a big deal? And anyway, didn't we get rid of most of them a long time ago, after the end of the Cold War and collapse of the big bad Soviet Union?

The answer to that last question is, of course, no. As of this writing, there are still approximately fourteen thousand nuclear weapons in the world, in the hands of the nine declared nuclear powers. That's far fewer than the maximum of between sixty to seventy thousand that existed at the height of the Cold War, but still a significant number.[1] Still, if we've managed to eliminate so many of them, then we can't have that much to worry about, right?

Unfortunately, that's also wrong. Nuclear weapons *are* a big deal, and that fact doesn't change even if the numbers of them do. The reason is that they *are* different, in profound and fundamental ways. Rather than simply "big bombs" scaled up to enormous dimensions, they work far differently than what's known as "conventional" weapons—a distinction that was completely superfluous and unnecessary before the first atomic bomb in 1945—and are far more powerful and dangerous as a result.

The Basics

In one sense, the "bigger bombs" idea is quite correct. There's no question that nuclear weapons are far more powerful than any other type of explosive, even if that power is commonly expressed in terms of one that's been around for almost two centuries: TNT (more scientifically known by its chemical name, trinitrotoluene). Because nuclear weapons were so much more powerful by orders of magnitude over anything previously known, much of the terminology around them had to be invented, based on concepts that had already been around for a while. Hence the practice of classifying nuclear weapons in terms of kilotons (thousands of tons of TNT) or megatons (millions of tons of TNT).

But that's only a convenient shorthand. The power of a nuclear weapon doesn't come from the same source as does the power of TNT. Conventional explosives such as TNT, dynamite, and nitroglycerin, among others, are all expressions of chemistry: they create violent energy through rapid chemical reactions. These reactions involve the electron bonds that enable atoms of different elements to join together in molecules and the various processes that can break apart those bonds. Such processes generally involve only the outer regions of atoms—that is, the electrons that dance around the nucleus at different energy levels—and not the nucleus itself. When chemical bonds are torn asunder, whether in the sudden violence of an explosion or more gentle means of other chemical reactions, the elements involved remain the same, even if they may subsequently recombine to form new substances.

Nuclear reactions, as the term implies, involve the nucleus of the atom: the protons and neutrons at its heart. The number of protons and neutrons determines the identity of the element, with the number of protons (known as the atomic number) determining the number of electrons too. Most familiar elements, such as oxygen, iron, or silicon, are stable, meaning that they contain an equal number of protons and neutrons in the nucleus. Other heavy elements with large nuclei, however, are unstable, with an unbalanced ratio of protons to neutrons, making the nucleus a restless, churning mass that continually emits particles, trying to restore balance. These, such as uranium, the most famous example, are known as radioactive elements, and these are what make nuclear energy—and nuclear explosions—possible.

Because such nuclei are inherently unstable, they will eventually decay into other, more stable elements. In most naturally occurring elements, that's a process that normally takes a great deal of time, from thousands to even millions of years. But that also means it doesn't take much to send them over the edge. It's possible to greatly speed up that process by doing nothing more than sending a neutron to collide with the nucleus. When that happens, in a process called "fission," the nucleus is split apart. The fragments of the former nucleus, each now a separate entity with a certain number of protons and neutrons, are new elements, and some of the neutrons spray off on their own into space. If other unstable nuclei are close enough, those stray neutrons can hit and tear them apart as well, with the process continuing and accelerating, releasing more and more energy in what's known as a chain reaction. If it all happens fast enough, on a timescale of less than a microsecond (one-millionth of a second), the result is a nuclear explosion, with the release of vast amounts of heat, kinetic energy, and radiation.

There's another way to get a nuclear explosion, however. Instead of splitting apart heavy nuclei, we can fuse together the lightest one of all, that of the hydrogen atom, consisting of a single proton. This also creates new elements, heavier this time, and releases great energy. Known as fusion, it's the fundamental process that makes the stars shine, including our own sun. You can experience the power of fusion for yourself every day; just go outside to see the sunlight and feel the warmth on your skin, and realize that it's coming from about ninety-three million miles away.

The problem with nuclear fusion, though, is that it requires extremely high temperatures, on the order of millions of degrees. That's because hydrogen nuclei—that is, protons—naturally repel each other with great force, and must be moving extremely rapidly to overcome that natural antipathy and fuse together. In the nuclear weapons business, this means that a fusion or hydrogen bomb needs a fission bomb to set it off because that's the only way such high temperatures can be instantaneously generated. In a way, then, every H-bomb is really two bombs in one. I will delve further into such aspects of weapon design as well as precisely what happens when a nuclear explosion occurs in chapter 5 when I discuss more of the technical details.

For now, the main points to understand are that nuclear weapons are fundamentally different in many ways from other explosive devices (whether one chooses to call those weapons conventional or some other comforting name) and come in two basic varieties: fission and fusion. The former type is what has traditionally been known as an "atomic" bomb, while the latter is termed a "hydrogen" or "H-bomb," or more scientifically, a thermonuclear weapon. Such terms are really only a rough approximation of the technical realities; while they both involve atoms, more specifically their nuclei, each type works differently. In general usage, however, most people don't really bother to distinguish between them, and consider atomic bombs, hydrogen bombs, and nukes all pretty much the same thing.

But not long after Hiroshima, the atomic bomb's special status in the public consciousness was signified by a typographic distinction that set it apart from its comparatively milder explosive predecessors. In countless newspaper articles, magazine pieces, and editorials, it became known simply as "the Bomb." No further explanation was necessary; everyone knew that the capital *B* meant the atomic bomb, not your average five-hundred-pound bomb dropped from the belly of an airplane or an artillery shell lobbed across a few miles. Only a nuclear weapon, no matter whether it was atomic or hydrogen by type, merited that capital letter and specific article. Those literary distinctions separated the past from an uncertain and frightening future.

Distinctions without a Difference

The "atomic" bomb came first, in 1945. The world saw three atomic explosions that fateful year, the first of them a test in the desert of New Mexico, and the second and third over Japanese cities. The hydrogen bomb came along in 1952, with the United States testing the first example by obliterating a small Pacific island. Since then, a dizzying variety have been devised and built, in all sizes, shapes,

In countless newspaper articles, magazine pieces, and editorials, it became known simply as "the Bomb." No further explanation was necessary.

and destructive yields. Although to date, only 2 nuclear weapons have ever been used in warfare—on Hiroshima and Nagasaki—about 2,041 of them have been exploded since then in scientific and military tests, which have been conducted in every conceivable environment and condition, including on, above, and below the ground; underwater, deep and shallow; every level of the earth's atmosphere; and hundreds of miles into outer space.[2] The weapons have been as small as suitcases and as large as houses. The explosions that they have generated have been as small as the equivalent of a few hundred pounds of TNT to over fifty million tons in the most powerful test ever conducted—the USSR's "Tsar Bomba" in 1961.

But the qualities that make nuclear weapons a different and separate category on their own are more than matters of a scientific, technical, and physical nature. Physicist Albert Einstein, who as we will see was one of the seminal figures whose work made nuclear weapons a reality, once remarked that the atomic bomb had changed everything, except our way of thinking. He was right, but only to a certain point. The prospect of the destruction of human civilization that became a practical reality with nuclear weapons certainly seems to have done little to eliminate our persistent drives for aggression, dominance, and warfare (though, as will be discussed later, it may have restrained them), yet since 1945, it has affected international politics, military strategy, and human psychology on collective as well as individual scales.

It's true that much the same can be said of other major developments in military technology, from crossbows and catapults to machine guns and land mines. Each of these weapons changed the conduct of warfare at a profound level in ways that not only affected the individual soldier on the battlefield but also reverberated all the way up to the leaders who were waging the wars and making the big strategic decisions. Unfortunately for the individual soldier, the true meaning of those changes usually isn't immediately apparent to the commanders responsible for their lives who might be ordering them to charge en masse into machine-gun fire or ignore the strange-looking clouds rolling over their trenches, without recognizing the dangers. After an inevitable costly learning period, however, military strategists learn the capabilities of the new, revolutionary weapons, and devise defensive measures to circumvent them or at least blunt their impact.

Such has not been the case with nuclear weapons. They've only been used twice in warfare. And because the second bombing, on Nagasaki, occurred merely three days after the first on Hiroshima, there was practically no time in between for the Japanese to begin to comprehend what that meant and how to respond.

Decades of nuclear testing have resulted in detailed studies of weapons effects and the development of various measures to protect human beings from them, such as fallout shelters, iodine pills, and evacuation schemes. Some have even been tested in elaborate simulation exercises. But will they actually work if the missiles fly? No one knows. We have practical experience in how to defend soldiers and civilians against bullets, shells, mines, chemical attacks, and pretty much anything else that can be thrown against them. That experience was earned the hard way, in actual warfare. It's sorely lacking with nuclear weapons.

That, of course, is because there's never yet been a real nuclear war—something for which we should all be grateful. And it brings us back to the factor that sets nuclear weapons apart from all others: they really *are* bigger bombs. Because the destruction that they can wreak is so much more than any other weapon, because they can, at a single blow, kill more people than anything else humans have devised and built, they simply have to be considered on their own terms.

Since nuclear weapons were created, suggestions have been advanced to tame them by using them for some kind of constructive rather than destructive purposes. One idea seriously considered during the late 1950s was called Project Plowshare, which proposed using atomic bombs essentially as landscaping tools for tasks such as leveling mountains, creating artificial harbors, or excavating huge holes. A few small-scale tests of the idea were even conducted, proving what everyone already knew, namely that

Because the destruction that they can wreak is so much more than any other weapon, . . . [nuclear weapons] simply have to be considered on their own terms.

atomic bombs could easily do such jobs. Unfortunately there was one glitch, namely the creation of tons and tons of radioactive dirt that somehow had to be disposed of, so Project Plowshare failed to progress beyond its grandiose impracticality.

Another fanciful plan was Project Orion, which aimed to solve several problems at once by using hundreds of nuclear explosions to propel a giant spacecraft, eliminating atomic stockpiles while undertaking an interstellar voyage. It was an inspired and ingeniously engineered idea from some of the world's most brilliant minds, but even more outlandish than Plowshare.

Such schemes may have served a useful purpose for those actively engaged in the creation of nuclear weapons, distracting them from the ultimate purpose for the existence of the devices. But only temporarily. The scientists and administrators of the Manhattan Project had not been trying to explore the cosmos or come up with new earth-moving techniques. They were desperately trying to build a new type of weapon, and to do so before the Nazis got there first. After the war, they were striving to preserve US supremacy, believing that the Soviet Union, like the Nazis before it, was intending to take over the world. In both cases, they were talking about weapons for war, not tools for peace—even if, during the Cold War and beyond, some tried to convince the world that nuclear weapons *were* actually tools for peace.

The singularly destructive power of nuclear weapons leads to another of their unique attributes—one that is simultaneously physical and practical, and psychological and philosophical: their identity as "doomsday" weapons. The concept of some kind of world-destroying power isn't anything new; it's been found in the religions and folklore of practically every culture since the beginnings of human civilization. But we never seriously possessed anything like it, not until the atomic and especially hydrogen bombs were invented.

In the immediate aftermath of Hiroshima and Nagasaki, after the new and unfamiliar concept of the atomic bomb had begun to sink into the consciousness of the public, a comforting belief started to take hold. The atomic bomb had ended the war because it was so powerful that no one could stand against it. Surely it would make war obsolete, at least on the scale of the full-fledged world war that the world had just survived. War would simply be too awful, too destructive, for any nation to contemplate it anymore, no matter the purpose. Instead, the atomic bomb would usher in a new era of international unity and peace, under the aegis of a global government, which would control and administer the awesome power of the atom for the good of all.

It was a glorious vision, but human nature and political realities being what they are, it was not to be realized. The United States enjoyed a brief period of exclusive global dominance as the sole possessor of the atomic bomb, aided by the fact that it had suffered far less damage in World War II than any of the other major combatants. That idyllic interlude ended abruptly in 1949 when the Soviet Union got the atomic bomb—a shock to which the United States responded by rushing ahead to the next step, achieving the hydrogen bomb in 1952. When the Soviets responded in kind less than a year later, it was clear that instead of realizing any dreams of international unity, nuclear weapons had brought about a dangerous age of anxiety on a level never before seen.

Central to that anxiety was the possibility of doomsday. For the average person, that didn't have to mean the literal end of the world, with the actual death of every single person on the planet. Just your own death, along with all your loved ones and the obliteration of your hometown, added to the devastation and collapse of your country's society, was doomsday enough. As the United States and USSR continued to increase their arsenals and threaten each other throughout the 1950s, as movies and TV shows and novels offered ever more frightening scenarios, and as governments began to think about ways to supposedly protect their citizens in the event of war, the prospect of doomsday became a real and ever-present cloud over everyone's heads, at least in the Northern Hemisphere, which would bear the brunt of any likely nuclear war.

What is different, then as well as now, is that the cloud hangs over everyone, utterly regardless of politics, ideology, ethnic identity, or any of the other labels that humans use to divide themselves. The possibility of nuclear war unifies the human race, if not in the way that the postwar idealists would have preferred. The reason is that it doesn't discriminate. Nuclear weapons are the ultimate equalizer: where one goes off, it kills and destroys everyone and everything within range. A firearm can be precisely aimed to kill a single human being; an artillery shell or drone can be exactingly aimed to destroy a single building or installation. For nuclear weapons, the smallest target is an enormous structure, town, or city. They are, by their very nature, quite impersonal, and at their largest scale, essentially genocidal.

Mutually Assuring Destruction

Although America may have felt profoundly threatened after the Soviet Union acquired the atomic bomb, and despite all the hysterical rhetoric and propaganda spouted by generals and politicians throughout the 1950s, it wasn't until the mid- to late 1960s that the United States really had cause to worry. By that time, the Soviets had acquired a rough nuclear "parity" or equality in nuclear weapons. The era of mutual assured destruction (MAD) had arrived.

MAD was in effect an official admission by the United States that what were now called the "superpowers"—that is, the United States and USSR—were tacitly participating in a mutual suicide pact. If each side had achieved the capability to destroy the other side as a functioning society, then neither could do so because retaliation in kind would also completely destroy the attacker.

So went the logic, which was essentially sound, assuming that rationality prevailed on both sides. But neither the United States nor the USSR was willing to accept the status quo. It was not enough, argued strategists and theorists, to settle for parity and the supposed stability it offered. Instead, each side had to be superior to the other, leading to an ever-intensifying arms race, meaning not only more powerful weapons but better ways to counter them too, coupled with ever-greater expenses that strained the economies of both nations.

The cost of maintaining huge nuclear arsenals, not to mention upgrading and expanding them in an endless upward spiral, did have beneficial effects. It led in the 1960s and 1970s to the first serious efforts at arms control, which although not entirely successful and often little more than an opportunity for political grandstanding, kept the superpowers talking and trying to understand each other. It also helped the United States and USSR to seem slightly less hypocritical in their reluctance to allow other countries entry into what had become known as the "nuclear club." The United Kingdom joined that exclusive group in 1952, with France following in 1960, both of which were more concerning to the Soviets than to the US government. But when China got the Bomb in 1964, the United States began to worry more about the "international proliferation" of nuclear weapons and how to encourage diplomatic efforts to limit it.

The reasoning of MAD also meant that if wars were still going to be fought, alternatives had to be found—which rather than some kind of new weapons and warfare, meant returning to the tried and true. If the superpowers couldn't fight each other directly, then they would do so by proxy, whether militarily through client states in remote parts of the world such as Korea and Vietnam, or covertly through espionage and shadowy schemes to challenge each other's influence. Europe was a special case because the USSR held unquestioned military superiority on that continent. The specter of Red Army divisions overrunning and seizing all of Western Europe haunted the United States from the end of World War II, even before the Cold War truly began (whether one dates that from the Berlin crisis of 1948 or Soviet acquisition of the Bomb in 1949). It was the major impetus for the formation of the North Atlantic Treaty Organization (NATO) in 1949 that united the United States with Britain and Western Europe to hold back Soviet expansion. The admission of the United States and NATO that its conventional forces couldn't possibly withstand Soviet invasion turned Western Europe into a nuclear trip wire, with the understanding that if Soviet

If wars were still going to be fought, . . . [and] the superpowers couldn't fight each other directly, then they would do so by proxy.

tanks ever began rolling into West Germany, they would be met with a nuclear response.

In theory, such a response would be only "tactical," meaning the use of small nukes on a local scale. But no one was really confident that a supposedly "limited" nuclear war could be prevented from quickly escalating into all-out global conflict. For NATO and the United States, that doubt, as scary as it might have been, was thought to be another benefit of deterrence as well as a reason to spend more money on conventional weapons, not only to keep troops stationed in Europe, but by the 1960s, to support the apparently interminable Vietnam War.

Yet the expense and political pressures of MAD, not to mention its ethically questionable foundations, compelled the consideration of more radical alternatives, such as the idea that just maybe, nuclear war *didn't* have to mean the end of the world as we know it. Perhaps nuclear war *wasn't* an all-or-nothing proposition, as had pretty much been the dominant belief from the start of the atomic age. Especially after the close call of the 1962 Cuban missile crisis and several similar incidents (which I will examine later), some strategists, generals, and politicians began to offer the heretical notion that it might actually be possible to fight a limited nuclear war and even win it.

The idea started to seem more practical given the technological advances in nuclear weapons and the various systems that delivered them to their targets. Beginning with the advent of intercontinental ballistic missiles (ICBMs) in the late 1950s, more options presented themselves. Instead of simply dropping nukes from bomber aircraft, we could now hurl them across outer space to enemy territory from the safety of our own borders or launch them from undetectable submarines off the coast of our enemy. If the Russians invaded Europe, we could hit them on the front lines with nuclear artillery. And now we didn't have to use multimegaton warheads that wiped out whole cities; we had weapons with yields at all levels. We could, said the nuclear planners, tailor a nuclear response to a specific attack. The president didn't have to choose between Armageddon and complete surrender.

It sounded attractive on the surface. But it also assumed that our enemy would respond according to our own preferences and expectations. When warheads began exploding over military bases, cities, or both, who really knew what would happen? If we did our adversary the courtesy of only nuking military bases while avoiding their cities, what guarantee was there that they would respond in kind, particularly given that by their very nature, the effects of a nuclear weapon, even a relatively small one, couldn't be limited to a preferred geographic area? If the Soviets, for example, were good enough to limit an attack to our ICBM bases in North Dakota, Montana, and Wyoming, the fallout would still threaten Chicago, Minneapolis, Saint Louis, Detroit, and all the civilian population downwind.

Still, plans were made, scenarios were devised, and weapons were deployed to make them possible. Military strategists and politicians comforted themselves with the belief that they had reasonable options. One of these, starting in the early days of the nuclear age and peaking in the 1960s, was the idea of civil defense—the hope that by building a network of public and private shelters in school basements and suburban backyards, and teaching school-children to dive under desks, a significant number of Americans could ride out a nuclear attack. The Soviets did much the same, especially in their largest population cen-ters. Even as experts pointed out the somber truth that with the advent of hydrogen weapons, most people cower-ing in shelters would simply be baked or suffocated inside by the firestorms raging outside, others argued that shel-ters farther away from the fury of direct nuclear detona-tions would protect people from the lingering fallout that would follow.

It was all little more than elaborate propaganda, a form of theatrical busywork to provide reassurance to the public of a nonexistent safety and security, or the late twentieth-century equivalent to the early twenty-first-century ritual of being forced to remove one's shoes and submit to ham-handed inspections before boarding an air-liner. When it became clear that it would be impossibly expensive to build enough shelters to protect more than a handful of people, official efforts for civil defense devolved into little more than the publication of information pamphlets and designation across the country of various public buildings as fallout shelters, some of which were stocked with a cache of basic food and medical supplies that eventually fell into unusability, and were never restocked. Most citizens soon realized, if they'd ever thought otherwise, that there was really very little they could do. If war came, they were on their own.

The Power of Perception

When we speak of nuclear weapons, we naturally discuss their explosive power, the incredible heat they create, the radiation they emit, and the fallout they leave behind. But as may have already become evident through my exploration of doomsday, MAD, and the difference between conventional and nuclear bombs, another quite powerful facet of their exceptionalism has nothing to do with physics or the technicalities of military strategy. Instead, it's based in human psychology, specifically the beliefs, perceptions, and even taboos and mythologies that surround anything nuclear.

The theory of deterrence, which I briefly mentioned above and will examine in greater depth in chapter 6, is a prime example. Hesitating to attack a nation that you believe to be your mortal enemy because you are convinced that doing so means your own destruction sounds logical—a classic no-win proposition—but for it to prevent war, you have to also believe a few other things. First, you have to believe that your enemy actually possesses that much power; second, that they won't hesitate to use it; and third, that they won't feel so threatened that they decide to take the chance of hitting you first. These are all matters of perception, threat and counterthreat, bluff and counterbluff, trying to figure out your enemy's true intentions, and even your own.

As we'll see later, the Cuban missile crisis of 1962 was the closest that the world ever came to actual nuclear war. It was a series of moves and countermoves, each proceeding to a more threatening level, almost to the point where the United States or Soviet Union felt it had no choice but to go to war. In the end, though, when staring that choice directly in the face and seeing nothing beyond but oblivion, both President John F. Kennedy and Chairman Nikita Khrushchev realized that ultimately, they *did* have a choice and resisted all the pressures trying to convince them otherwise. And they stepped back from the brink. Part of that was due to sheer luck, yet it was also because we were fortunate enough to have two world leaders in control who were smart, sensitive, and human enough to weigh the true costs of taking the final step into war, and decided that it wasn't worth it.

In Stanley Kubrick's classic film *Dr. Strangelove*, the title character, a nuclear strategist for the Pentagon, observes that "deterrence is the art of producing in the mind of the enemy the fear to attack." No doubt, fear on a very human level was certainly a factor in staying the hand of Kennedy and Khrushchev in 1962, and other world leaders in other crises. But human empathy and compassion as well as a cool risk versus benefit calculation figured into the outcome. And that's true when both considering the positive (so far) outcome of every nuclear confrontation that has yet occurred and asking the question that many have pondered when wondering how we've (so far) managed to go seventy-five years since the last time a nuclear weapon has been used in anger. Has some sort of unspoken moral taboo arisen against nuclear weapons that's saved us, that's always managed to keep the president or premier or prime minister in question from pressing the fateful button at the last possible moment?

There may be no way to adequately answer that question. But just posing it to begin with once more highlights the fact that however one may choose to define or explain it, nuclear weapons are fundamentally different from other weapons. The remainder of this book will explore the various aspects of those differences and what they mean, starting with a look at how they came into existence in the first place.

DAWN OVER TRINITY:
THE NUCLEAR AGE BEGINS

When the atomic bomb became a reality in 1945, the world responded with shock, fear, and wonder in equal parts. Yet the Bomb did not come as a complete surprise. It's true that thanks to the extreme secrecy of the Manhattan Project that built the weapon, the US public had absolutely no idea that approximately $2 billion of its tax dollars had gone into building an atomic bomb (though as we will see, other parties, notably the Soviet Union, had been quite aware of it). The idea of some kind of superweapon having something to do with atomic something or other had been around for quite awhile, though, in comic strips and pulp science fiction magazines and on radio programs.

Soon after a number of scientists began to notice and investigate a mysterious natural phenomenon that became known as radioactivity, and discern and study the inner geography and lives of the atom, ingenious and scientifically savvy fictioneers picked up on the new discoveries and ran with them, unleashing their imaginations to speculate on the potentialities, both good and bad. H. G. Wells, already world famous for inventing time machines (in *The Time Machine*, 1895) and masterminding invasions from Mars (in *The War of the Worlds*, 1898) among other wonders, was not only a trained scientist but also a trenchant social critic, fascinated by politics as well as the means by which humans tried to relate to each other, fight among themselves, and govern themselves.

So it was only natural that in 1914, he would publish a novel, of the sort he liked to call "scientific romances," about global warfare using super scientific weapons. Wells's *The World Set Free* (1914) contains what appears to be the first use of the phrase "atomic bomb" along with the first depiction of atomic bombs as weapons. In the novel, directly inspired by and dedicated to the scientists who had discovered radioactivity less than two decades before, cities are destroyed by atomic bombs dropped by aircraft. The resulting near destruction of civilization leads to the birth of a new world with freedom and peaceful atomic power for all.

Wells would live until 1946—long enough to witness the partial realization of his darker visions over Hiroshima and Nagasaki, if not his more optimistic hopes of a world government guided by science and rationality. But the scientists and engineers who created nuclear weapons weren't trying to fulfill the predictions of a visionary

British writer. At least initially, they were attempting to see whether such a weapon was truly possible, and if so, develop it before the Nazis could use it to enslave the world. When the first thin threads of theory and experiment that ultimately led to the Bomb were picked up and followed, beginning almost fifty years prior to Hiroshima, no one quite knew where they would lead or how convoluted the path would be.

The Secrets of Creation

The man to whom Wells dedicated *The World Set Free* was Frederick Soddy, a brilliant young chemist who, collaborating with the even more brilliant physicist Ernest Rutherford, had proven that radioactivity, which had only been discovered several years earlier in 1896 by physicist Henri Becquerel, involved certain elements such as radium somehow changing or "transmuting" into other elements. The specific details of how that happened remained to be worked out, but it was already clear that in some way, a great amount of energy was released in the process. If some means of controlling and channeling that energy could be found, proclaimed Soddy and others, the atom might provide an endless source of power for humankind—or also a weapon. Wells went on to incorporate both choices in his novel.

Rutherford had discovered the nucleus of the atom and two of the types of radiation that it could emit—alpha, beta, and gamma—but what was going on inside the atom was still not completely apparent. The new quantum theories of physics were revolutionizing the field, and demonstrating that it was all far more subtle and complex than anyone had previously suspected. Einstein provided another key to atomic energy with his theory of mass energy equivalence, as expressed in his famous equation $E = mc^2$, showing that even a tiny amount of mass contained a vast amount of energy, if it could only be released.

When physicist James Chadwick discovered the neutron in 1932, he found the final necessary piece. Now there was a way to penetrate the barrier of the positively charged nucleus—something that the negatively charged electrons or positively charged alpha particles that had been tried previously could not do. What would happen if someone tried shooting neutrons into the nuclei of heavy elements such as uranium?

In 1934, Italian physicist Enrico Fermi began a series of experiments to answer that question, methodically bombarding almost every element in the periodic table with neutrons. When he got to uranium, known as the heaviest naturally occurring element, things got interesting, including what appeared to be the inexplicable production of lighter elements. Later experiments by chemists Otto

Hahn and Fritz Strassman at the Kaiser Wilhelm Institute in Berlin were analyzed by physicist Lise Meitner, Hahn's brilliant colleague. Meitner realized the answer, so simple yet so elusive because no one had believed it possible: the uranium nucleus had split into fragments, each obviously consisting of fewer protons, making them lighter elements. Hahn and Strassman had actually split the uranium nucleus into two near-equal halves, in the process releasing enormous amounts of energy. Borrowing a term from biology, the phenomenon was called "fission," equating the splitting of the nucleus of the atom to the division of a cell nucleus.

Danish physicist Niels Bohr carried the news to the United States, sharing it with his colleagues there, including physicists Leo Szilard and Fermi, who now realized that he had actually split the atom and not even known it. A minor design quirk of his experimental setup had obscured his results and delayed a major scientific discovery.

Although nuclear fission was immediately recognized as a monumentally significant breakthrough in physics, not many people were worried about atomic bombs yet. Fission was obviously a phenomenon of great scientific importance and interest, but whether it could be harnessed for practical civilian or military purposes remained an open question. Some other particularly visionary individuals, however, weren't quite so sanguine. One of them was Szilard, a Hungarian refugee who by 1939 had temporarily settled at Columbia University in New York City. Szilard was the type of intellect who was generally several steps ahead of most everyone else, even including his scientific colleagues.

He had been thinking about atomic energy, including bombs, longer and more intensively than most. Quite familiar not only with Wells's *The World Set Free* but also with Nazi tyranny, Szilard was quite disturbed by the fact that fission had been discovered, of all places, in the heart of Nazi Germany. What if Adolf Hitler heard about it, realized its potential, and devoted all of Germany's renowned scientific prowess—even if that had been greatly depleted by the exile of many of its top minds, including Einstein—to researching and perhaps even building an atomic bomb?

In spring 1939, close on the heels of the announcement of nuclear fission, Szilard conducted an experiment at Columbia that confirmed his worst fears. Uranium did indeed appear to produce enough extra neutrons to sustain a chain reaction. Similar experiments by others, including Fermi, gave identical results. Szilard recalled later that at the time, "there was very little doubt in my mind that the world was headed for grief."[1]

Confirmation of a chain reaction also confirmed the possibility of a bomb, at least theoretically; the practical details of building such a device were still a mystery.

Although nuclear fission was immediately recognized as a monumentally significant breakthrough in physics, not many people were worried about atomic bombs yet.

But just the possibility was enough for Szilard, thinking of such a bomb in the hands of the Nazis. While he and other colleagues continued their investigations of fission, neutrons, and what particular isotopes of uranium might best lend themselves to a chain reaction, he decided that the United States and other free governments of the world weren't taking matters seriously enough, and set out to convince them. They not only had to take steps to protect the world's sources of uranium ore, located chiefly in Belgium's African colony and Czechoslovakia (now under Hitler's control), but also be made aware that such a bomb was even possible.

Szilard realized that his fame and influence outside scientific circles was fairly limited, so he enlisted the aid of the world's most famous scientist, Einstein, who agreed to place his illustrious signature on a letter to President Franklin Delano Roosevelt dated August 2, 1939. Szilard kept the letter restrained and matter of fact, informing FDR that recent scientific work had revealed the probability of creating "extremely powerful bombs of a new type" through "a nuclear chain reaction in a large mass of uranium."[2] The letter suggested that some kind of permanent contact should be established between the government and scientific community, and perhaps some funds could be made available to speed up the scientific work. Also mentioned were the uranium supplies in the Belgian Congo and now Nazi-dominated Czechoslovakia.

Interestingly, as much as these items may have been news to Roosevelt, they were likewise to Einstein. So involved in his own wholly theoretical and abstract work, comfortably settled into pleasant isolation at the Institute for Advanced Study in Princeton, New Jersey, since arriving from Germany six years previously, he hadn't been keeping up with the latest developments in physics. When Szilard filled him in on the news about fission and chain reactions, Einstein responded, "I never thought of that!" Although he'd certainly known of the energy locked inside the atom, only a few months earlier, Einstein had confidently told *New York Times* science writer William L. Laurence that it would remain inaccessible by humans for the foreseeable future, saying, "We are poor marksmen, shooting at birds in the dark in a country where there are very few birds."[3] Still, Szilard recalled, he was "very quick to see the implications."[4]

Unfortunately, late summer 1939 turned out to be a bad time to be writing letters to FDR. On September 1, Germany invaded Poland, plunging humanity into World War II. Roosevelt's time was completely occupied trying to find ways to help America's allies. It wasn't until mid-October that the letter finally reached the president.

Roosevelt got the point immediately. "What you're after is to see that the Nazis don't blow us up," he told the military aide that brought him the letter. Declaring that "this requires action," he directed the establishment of a committee to investigate the uranium question further. Yet not much happened for some time in official US circles, even as the British and Soviets began to investigate the prospects for an atomic bomb.

Throughout 1940 and 1941, slowly but steadily, a momentum developed. In mid-1940, the United States established the National Defense Research Committee under MIT scientist Vannevar Bush to begin marshaling US scientific resources for war, including the ongoing research on uranium. A year later, Bush became the director of an organization with broader powers and authority, the Office of Scientific Research and Development, with Harvard president James Conant taking over the National Defense Research Committee. An official British report stated that "we have now reached the conclusion that it will be possible to make an effective uranium bomb," and advised continuing and extending Britain's collaboration with the United States. Still the United States dragged its heels. It was not yet actually at war, after all.

That changed on December 7, 1941, when Japan attacked Pearl Harbor. The United States declared war on Japan, and Germany and Italy did likewise on the United States. On January 19, 1942, Roosevelt officially authorized the atomic bomb project, with Britain as a collaborator. In August, the effort received its official designation: the Manhattan Engineering District, or as it became more commonly known, the Manhattan Project.

The Path to the Bomb

As the various scientific groups and technical studies scattered all across the United States and Britain began to come together under the official Manhattan Project umbrella, they faced some pressing immediate questions. It was clear that scientifically speaking, building an atomic bomb was certainly *possible*. But no one actually knew how to do it.

There were many pieces to the puzzle, including the particular uranium isotopes that would be needed, how to start the nuclear reaction, and how to build a bomb that wouldn't merely fizzle out before actually exploding. The first issue was paramount. The scientific work over the previous several years had already established that naturally occurring uranium, uranium-238, wouldn't work for a bomb; its much rarer isotope, uranium-235, would be needed. But natural uranium contains less than 1 percent of U-235, meaning that some means had to be devised for separating and concentrating it from uranium ore.

It was a far more complicated problem than it sounds. Separating different elements from one another is a relatively simple chemical process, based on their different behaviors and chemical characteristics under certain conditions. Separating two isotopes of the same element is much harder, however, because they're essentially the same except for a minuscule difference in mass. The only

It was clear that scientifically speaking, building an atomic bomb was certainly *possible*. But no one actually knew how to do it.

difference between U-238 and U-235 is three neutrons, a matter of mass, not chemistry.

There were various ideas for how to do it with uranium, but no one knew which would work or how well. One idea was called gaseous diffusion, in which uranium is converted to a gaseous form that is forced through a porous membrane through which only the lighter isotope can pass. Another was electromagnetic separation, in which particles are accelerated in a cyclotron under a magnetic field, causing the different isotopes to move in slightly different paths. Other methods existed, but whichever was chosen, it was clear that each of the separation strategies would require huge, complex, expensive industrial plants. Figuring out how to build them and where would be one of the major parts of the Manhattan Project.

Another possibility had recently emerged from the uranium work at the University of California at Berkeley. Using the university's cyclotron to bombard uranium samples, chemist Glenn Seaborg had discovered a new stable element beyond uranium, number 94, which would eventually be named plutonium (because it was the next element beyond element 93, named neptunium). Experiments with plutonium showed that it could be made to fission, although it wasn't yet clear how it could work in a bomb. Still, plutonium promised an alternative to uranium, so it too would be pursued.

Whether an atomic bomb was going to use uranium or plutonium, the next question was the actual, physical design of a bomb. How would it actually work? What would it be made of? How big would it be? Would an airplane be able to carry it, or would it have to be delivered to its target by a ship? These were questions more for engineers than scientists, but they would have to work closely together to find the answers.

The two men who would lead the Manhattan Project would represent both disciplines. Since the project was officially under the auspices of the Army Corps of Engineers, its chief administrator came from that department: Colonel Leslie R. Groves. A hard-nosed, gruff, talented civil engineer, Groves had just finished building the Pentagon, so he was a man greatly experienced in getting massive, complicated projects done.

Taking charge in September 1942, Groves immediately set to work. He bought land for an isotope separation plant in eastern Tennessee, a place called Oak Ridge, and other facilities around the country. He began securing supplies of uranium ore, and lining up the companies that would construct and operate project facilities. In the process, he also started to absorb all the immense amounts of information that he would need to know, consulting with scientists, engineers, and officials, gathering the resources that would be necessary for what was clearly going to be the largest and most ambitious engineering project of his entire career, if not all US history.

As Groves charged around the country getting everything moving, he began to realize that the whole effort not only needed far more organization but more centralization as well. There was a war on, after all, and time was of the essence. It wouldn't do to have scientists working at different labs all over the country, in New York and Chicago and California and wherever else, wasting time and perhaps duplicating each others' efforts. Also, there was the matter of security. How could the project be kept secret if its people were all over the place, possibly talking to people they shouldn't?

Obviously, a central laboratory would be the solution, located in some remote, isolated area where it could be protected and guarded, not only from nosy intruders getting inside, but from its resident personnel wandering too far afield and perhaps spreading secrets. While it wasn't the sole factor, the fact that Berkeley physicist J. Robert Oppenheimer shared a similar point of view was, in Groves's eyes, undoubtedly a strong argument for giving him the pivotal job of directing that laboratory.

The thirty-eight-year-old Oppenheimer, known to all as "Oppie," already had a connection to the atomic bomb effort. He was quite familiar with most of the key players in the whole fission story, having studied and worked closely with them during his student days in Europe, and had discussed its implications with them both professionally and personally. In summer 1942, he had led an informal colloquy called the Berkeley summer study to look at the details of how an atomic bomb might be designed. The participants didn't settle the issue, but they identified some interesting and important questions that needed to be explored further.

Although they were polar opposites in most fundamental ways, Groves and Oppenheimer somehow took to each other on their initial meeting in October 1942 when Groves visited Berkeley to check on the work being done there. Groves, newly promoted to brigadier general, liked the sophisticated physicist's wide-ranging expertise and particularly his practical attitude, which seemed quite at odds with the "prima donna" scientists that Groves had so far met on the project. From Oppenheimer's point of view, Groves was quite different in temperament from his scientific colleagues, but seemed like just the no-nonsense type who could readily pull together all the disparate threads of the bomb project and get it done as soon as possible.

Groves was a man confident in his own intuition and decided that he'd found the man to direct the scientists of the Manhattan Project. He overcame, ignored, or simply bulldozed through various objections, including that the Nobel Prize–less Oppenheimer would be supervising

Nobelist colleagues who might resent it, and the fact that he had no administrative experience at all aside from dealing with his graduate students. A more serious issue to some, especially the FBI and army security people, was that Oppenheimer had past political and personal connections with leftist and Communist elements (including his own brother and wife). Such prior indiscretions didn't matter, Groves insisted; Oppenheimer was absolutely indispensable to the project.

Groves got his way, and several weeks after that first meeting, Oppenheimer was officially named director of Site Y, as the central lab would be designated. Oppenheimer suggested a suitable site, atop a New Mexican mesa near Los Alamos, a region where he vacationed and knew well. Groves moved quickly to secure the land, and sent in army engineers and contractors to begin construction as Oppenheimer embarked on a cross-country campaign to recruit its inhabitants.

It was a true test of Oppenheimer's leadership skills and personal charisma, as he attempted to convince scientific colleagues and friends to drop whatever they were doing for the duration of the war to go live in some remote place that Oppie wasn't allowed to name to do some kind of vital work that Oppie wasn't allowed to describe. All he could tell them was that they would be contributing their skills to a project that might actually win the war. For many scientists made refugees by Nazi tyranny, that was

Figure 1   Though an unlikely pair, General Leslie R. Groves and J. Robert Oppenheimer came together to lead the Manhattan Project. (US Department of Energy)

enough incentive. And of course, many, if not most, were able to guess what they were being asked to do, already aware at least indirectly of the recent discoveries that had rocked the physics community—and the possibilities for a bomb. Few of the people that Oppenheimer approached turned him down, even if it meant uprooting their lives and families for an uncertain future.

Until the central lab at Los Alamos went into operation, though, the major scientific work centered on the University of California at Berkeley, where physicist Ernest Lawrence's cyclotrons were busily churning out minute samples of U-235 and element 94 for experiments, and at the Metallurgical Laboratory ("Met Lab") of the University of Chicago, where Fermi and Szilard had moved to complete work on achieving a controlled fission reaction by building the world's first nuclear reactor. Szilard had conceived of such a device years before, but had lacked the facilities or experimental abilities to actually realize it. Now the means were available, including the gifted experimentalist Fermi to bring everything together.

Today, we think of nuclear reactors as sources of electric power, but in 1942, they provided a way to create the material to fuel a bomb—mostly, the newly discovered element 94, plutonium. Lawrence's cyclotrons could make plutonium only in painfully tiny amounts, atom by atom, gram by gram. That was good enough for lab experiments, yet completely impractical for the large-scale production that would be required for a bomb, which would need pounds, not grams, of plutonium.

Fermi's reactor, or "pile" as he called it, was essentially a makeshift lab experiment scaled up to fill a large space under the squash courts at the University of Chicago's Stagg Field. On December 2, 1942, Fermi and his team achieved a controlled nuclear chain reaction for the first time. They celebrated by drinking some cheap Chianti out of paper cups.

"The event was not spectacular, no fuses burned, no lights flashed," Fermi recalled later. "But to us it meant that release of atomic energy on a large scale would be only a matter of time." Arthur H. Compton, the Nobel prize-winning physicist in charge of the Chicago division of the Manhattan Project, called Conant of the National Defense Research Committee to inform him, by prearranged code, that "the Italian navigator [Fermi] has just landed in the New World."[5]

Despite the fact that the day's events had vindicated the work he'd conceived about ten years earlier, Szilard found himself feeling more somber than triumphant, knowing that now there would be atomic bombs to possibly destroy the world before atomic power might save it. "I shook hands with Fermi and I said I thought this day would go down as a black day in the history of mankind," he remembered.[6]

Building the Gadget

With the establishment of huge industrial-scale plants in Oak Ridge, Tennessee, and Hanford, Washington, the bomb design lab in New Mexico, and several smaller facilities in Chicago and scattered across the country, the broad shape of the Manhattan Project was now established. While the Oak Ridge and Hanford facilities concentrated on making the fissile material for a bomb, Los Alamos focused on building it. Early on, for security reasons, Los Alamos personnel were ordered to refrain from any references to bombs and instead instructed to call the object of their work "the gadget."

That work eventually settled around two designs. The first and simplest was a "gun" design in which one piece of subcritical uranium-235 would be propelled like an artillery shell into another subcritical mass of U-235, creating a mass that would go supercritical and then explode. Though they faced various technical details and engineering problems to overcome, Oppenheimer and his team were quite confident of the essential theory of the gun design. As long as they had enough U-235, it would work.

The second design was implosion, an elegant idea on paper, but a much tougher proposition to turn into a practical reality. Calculations and deeper study on the properties of plutonium made it clear that it wouldn't work in the gun-type bomb and would only result in a messy radioactive "fizzle," not a nuclear explosion. The only way to induce plutonium fission fast enough for an explosion was to squeeze a subcritical amount into a much smaller and denser critical mass. That meant surrounding it on all sides by explosives that were detonated simultaneously, creating a uniform pressure wave.

But no one had ever done such a thing before or even knew whether it was possible. Los Alamos scientists had to create assemblies of fast and slow explosives that were carefully calibrated to literally focus explosive force in a precise direction and intensity, exactly as a lens focuses light—explosive lenses. Hundreds of tests with conventional explosives using metal pipes and spheres allowed the theory to be perfected through laborious trial and error.

Although it wasn't given much attention at the time in the frantic rush for the atomic bomb, another weapon preoccupied one of the geniuses at Los Alamos. The theoretical discussions on the atomic bomb had ranged far and wide, including the idea of a bomb based not on fission but rather on fusion. Calculations showed that such a bomb, dubbed a "super bomb" or just "the Super," would be far more powerful than a mere fission device by orders of magnitude, perhaps to unlimited degree.

Edward Teller, another of the Hungarian émigré scientist group that included Szilard and physicist Emilio Segré, latched fervently onto the concept of the Super and insisted that it had to be pursued, dismissing any arguments that such efforts would be futile because it wouldn't be possible to build a Super without first achieving the fission bomb. And while there were good, workable ideas of how to build a fission bomb, no one knew how to begin designing a Super. But Teller's dedication (some would say obsession) with the Super was such that he was emphatic about being allowed to work on it at Los Alamos rather than the more pressing fission weapon project. Oppenheimer let him have his way, but the issue was one factor in a growing rift that would later cause serious trouble for both men.

By summer and into fall 1943, the Los Alamos lab was fully staffed, and up and running. The work involved over five thousand people, men and women, civilian and military, and encompassed far more than nuclear physics. All the research and work involved in building the gadget also required chemists, metallurgists, explosives experts, mathematicians, welders, clerks, cooks, drivers, military police, meteorologists, electricians, and practically every other scientific specialty and technical skill. Relatively few of the people at Los Alamos or any of the other Manhattan Project sites had any inkling of the big picture, nor were they encouraged to inquire further.

As work continued throughout 1944 and into 1945, the biggest concern facing Oppenheimer and Groves, apart from a stream of technical issues surrounding the bomb design, was whether there would be enough fissile material, whether uranium or plutonium, to actually build a practical weapon. The reactors at the Hanford plant were operating continuously, as were the gaseous diffusion and electromagnetic separation plants in Oak Ridge, yet only painfully miniscule amounts of U-235 and plutonium were being produced.

Meanwhile, US troops were dying every day in Europe, Nazi rockets were falling on London, and Japanese forces fought on ferociously in the Pacific. It was still unclear just what progress Nazi Germany or Imperial Japan might be making toward an atomic bomb, or whether in fact they were even making the effort. What was apparent was that even as the momentum of the war began to shift in favor of the Allies, it was far from over, and becoming bloodier with each passing month. The bomb had to be built, whatever it cost, and before anyone else had it.

For all the refugee scientists working on the Manhattan Project such as Szilard, Fermi, Teller, and many others, the defeat of Nazi Germany in May 1945 came as an immense relief. For some, it presented a dilemma too. The entire reason for the atomic bomb effort, the impetus behind Szilard's entreaties to the British and US governments, the motivating force that led Teller and Fermi and so many others to devote their entire energies to building an atomic weapon, had now vanished. An intelligence unit organized by Groves that followed advancing Allied armies throughout Europe into Germany to assess the progress

[The war] was far from over, and becoming bloodier with each passing month. The bomb had to be built, whatever it cost, and before anyone else had it.

of any Nazi atomic project had found nothing but a disorganized and crude effort, with scattered experimental attempts that led nowhere. Despite the worst fears of the refugee physicists and Allied governments, Hitler and the other Nazi leaders had failed to recognize the potential of the atomic bomb, and hadn't given it the official support it needed.[7]

So why, asked Szilard and a number of other scientists, mostly based at the Met Lab, should the effort to build a bomb continue? If it was no longer needed to counter a possible Nazi weapon, why not suspend the project and instead make some attempt to control other countries from acquiring a bomb? Why not concentrate now on nuclear energy for peaceful purposes as opposed to destruction?

Such sentiments were quite noble and perfectly understandable, but they were also completely futile. By this time, the Manhattan Project had taken on a momentum of its own, spurred by years of total war. When Roosevelt, the man who had set it into motion, died shortly before the end of the war in Europe and was succeeded by Harry S. Truman, that momentum only intensified. Truman had been totally unaware of the Manhattan Project, but once informed, saw no reason not to allow it to continue toward its culmination, especially as he looked ahead to the dark prospect of invading Japan.

At Los Alamos and the other project sites, the end of the European war hardly slowed down the work schedule.

Instead, preparations were underway to test the implosion weapon, for which enough plutonium would soon be available. With the gun design considered to be perfected, work had been concentrated on implosion. But enough uncertainty remained that Groves and Oppenheimer decided that the design would have to be tested before the weapon could be used—now not on Germany, but on Japan.

As Truman settled into office and prepared for his first meeting with his counterparts, Winston Churchill and Joseph Stalin, at Potsdam in Germany, a testing site was prepared in southern New Mexico on an army bombing range, a remote hardscrabble place in a desert known as the Jornada del Muerto (Journey of Death) near Alamogordo. The test device would sit atop a hundred-foot steel tower, surrounded by test equipment to measure and record every detail of the blast, along with observation points strategically placed miles away at what were hoped to be safe distances.

The plan had originally been to hold the test on July 4, with the obvious significance of that date, but technical problems squashed that idea. By the middle of July, everything was ready for the test shot, which Oppenheimer had named Trinity. As Truman and his advisers waited for the word, Groves and Oppenheimer faced their most anxious moment, with all their work, not to mention about $2 billion of taxpayer money, riding on the outcome of the shot.

As the test approached on July 15, the weather was not cooperating, and by the next day, conditions weren't looking much better, with heavy thunderstorms and rain predicted throughout the night. Conducting the test under less than perfect weather wasn't an option; aside from affecting the crucial measurements and observations of the explosion (if any), it could be dangerous because no one really knew if or how rain might carry radioactive fallout to populated areas. No one, in fact, really knew for certain whether the gadget would work at all, or if it did, how well. A betting pool was held on the size of the blast, with wagers ranging from complete destruction to a total dud.

Finally, in the wee hours of July 16, the meteorologists told Groves what he wanted to hear. There would be a period of clearing around dawn. Trinity would go on. Across the desert, final preparations were made.

Then, at 5:29:45 a.m., the final relays were closed and the Trinity gadget detonated in a brilliant glare never before seen on the earth. Reactions spanned the entire range of human emotions, from utter jubilation to deep satisfaction to complete awe to abject horror. Most of the scientists were simply relieved and happy that the goal for which they had worked so diligently for years had been unquestionably achieved. Somewhat later, many would think beyond their initial reactions to consider the deeper implications of what they had witnessed.

Figure 2 The completed Trinity device awaits testing atop the test tower in Alamogordo, New Mexico. (US government)

As befitting such a complex individual, Oppenheimer's reactions were ambiguous. Aside from the immediate satisfaction, he later recalled his thoughts on seeing the Trinity fireball—a quotation from the Hindu scripture: "Now I am become Death, the Shatterer of Worlds."

At the time, however, as a bystander recollected, Oppenheimer was too physically exhausted and emotionally spent to say anything other than, "It worked."

A Rain of Ruin

Events moved swiftly after Trinity. An ecstatic Groves cabled the news to Truman, who soon passed it on in a guarded form to Stalin, telling him that the United States now possessed a new weapon of "unusual destructive force." A seemingly unimpressed Stalin responded only that he hoped the United States would make good use of it against Japan. As Truman would learn much later, Stalin was already quite aware of atomic bombs, including the fact that the United States was building them. Despite all of Groves's elaborate security precautions, Soviet spies had infiltrated the Manhattan Project.

The first bomb, a gun-type bomb dubbed Little Boy because of its smooth, elongated shape, was dispatched to the Pacific in several pieces by ship and airplane, its uranium heart carried separately for safety reasons. Another weapon, identical to the plutonium Trinity device that had just proved itself in New Mexico, was also being prepared.

Ever the troublemaker, Szilard circulated a petition urging the president to consider the moral implications of using the bomb on Japan without warning. Although he collected over a hundred signatures from Manhattan Project personnel in Chicago and Tennessee, the petition was officially ignored. With the vicious Pacific war still raging, tensions growing between Russia and the Allies, an inevitably bloody invasion of the Japanese home islands planned in the fall, and a weary population anxious for any means to end the war, any questions of delaying or forgoing the use of a now-proven superweapon against the United States' remaining enemy were beyond serious discussion. In June, the Interim Committee along with a scientific subcommittee including Oppenheimer, Fermi, and Lawrence considered the idea of staging a demonstration of the bomb before using it, giving the Japanese a chance to see what awaited them. But the notion was rejected. What if the Japanese shot down the plane on its way to the demonstration? What if the bomb was a dud? What if the Japanese officials witnessing the test shrugged it off as a trick or refused to be suitably impressed?

Similar concerns surrounded a proposal to warn the Japanese beforehand. No, concluded the Interim Committee, the only course of action was to use the bomb by surprise, without any advance warning, on a previously untouched Japanese city that contained important military or industrial targets surrounded by workers' homes. A target committee picked several likely candidates, including Kyoto, Hiroshima, Yokohama, and Kokura. US secretary of war Henry Stimson vetoed Kyoto because of its unique historical and cultural importance.

Just after 8:15 in the morning on August 6, 1945, a B-29 Superfortress bomber named *Enola Gay* dropped the uranium gun bomb dubbed Little Boy over Hiroshima. It detonated 1,850 feet over the city, near the Aioi Bridge at the conjunction of the Honkawa and Motoyasu Rivers, with an explosive force of about fifteen thousand tons of TNT, or fifteen kilotons. The fireball of about three hundred thousand degrees Fahrenheit set the city afire, and the burst flattened everything within a two-mile radius. About eighty thousand people were killed immediately, whether from the blast, heat, or both.

The secret was out. The atomic bomb had been released on the world and into human consciousness. "The force from which the sun draws its power has been loosed against those who brought war to the Far East," declared President Truman, promising that "we shall completely destroy Japan's power to make war. . . . If they do not now accept our terms they may expect a rain of ruin from the air, the like of which has never been seen on this Earth."[8]

To emphasize that point, three days later, on August 9, another B-29, *Bock's Car*, dropped Fat Man, a plutonium implosion bomb identical to the one tested several weeks earlier in New Mexico. Though the bomb was somewhat off target and the early death toll was less than at Hiroshima partly because of the hilly geography of the city that confined the initial effects to a smaller area, the twenty-one-kiloton blast initially killed between thirty-five to forty thousand. Later that day, the USSR also declared war on Japan.

Still, stunned from the physical and psychological effects of the atomic bombings, and divided by internal conflict between civilian and military factions, Japan didn't surrender immediately. Preparations continued for a third atomic bombing, and perhaps a fourth if it proved necessary.

It did not. On August 14, 1945, the Japanese Empire surrendered unconditionally. World War II was finally over.

The atomic age, however, had just begun.

TAMING THE NUCLEAR GENIE: NEW HOPES AND THREATS

Suddenly, nuclear weapons were no longer the science-fictional fantasies of writers or the object of the dark fears of visionary scientists. The Bomb was a real, practical weapon of war that had killed thousands of people and, it was argued, prevented the deaths of thousands more by ending the worst war in history. At first, that latter assertion was all that most people cared about. The war was over, and if it had been the Bomb that ended it, then so much the better. Multitudes of American boys would now be coming home instead of perhaps dying on the Japanese home islands.

And the United States was on top as the only nation in the world with the secret of the Bomb. Of all the great powers, it had suffered the least in the war just ended and was now in a position to set the terms for everyone else in the postwar world. That fact wasn't lost on the president of the United States or his civilian and military advisers, nor on the United States' allies, such as Britain, or its rivals, such as the USSR.

But with that position of unquestioned supremacy came a set of unique opportunities and responsibilities. With the United States in sole possession of the supposed "secrets" of the atom, at least for the moment, there was a chance for it to ensure some kind of security for the future. The United States had been fortunate and prescient enough to get the Bomb first, but scientists warned that the same principles of physics couldn't be restrained by security classifications and top secret stamps. It would be only a matter of time before other nations discovered the same "secrets" of nature and eventually harnessed the atom for their own purposes, whether military or civilian. Inadvertently making that same point less than a week after Hiroshima, the United States officially published a report titled *Atomic Energy for Military Purposes*, also known as the Smyth Report after its author, Princeton physicist and Manhattan Project veteran Henry DeWolf Smyth. The report laid out in considerable detail the basic principles of the Bomb.

Even before Hiroshima, visionaries such as Szilard and Bohr tried to warn political leaders that some kind of international controls would have to be put into place to avoid an eventual atomic anarchy, a scenario of unrestrained competition leading to an arms race that would endanger the peace that had just been so expensively won. With the recent founding of the United Nations, the framework was in place to create a structure to, if not stuff the atomic genie back into its bottle, at least keep it from getting out of control and destroying the world.

As with most idealistic ventures, however, it would be far easier said than done.

Doomsday or Paradise?

One of the great problems in devising some means of controlling or regulating nuclear weapons in the immediate months and years after Hiroshima was trying to figure out just what the weapons meant, and in what context. For most military thinkers, the Bomb wasn't anything all that special, except in scale. It represented more destructive power for a smaller investment, with one bomb and one aircraft doing the work of hundreds of conventional bomber aircraft—in other words, a bigger bang for the buck. For others, including most of the scientists who had devised and built it, the Bomb was quite different; it was something that had never before existed in human awareness or capabilities. There was a fundamental, qualitative difference between atomic and conventional weapons (a distinction that now had to be routinely made). The ability to destroy an entire city and kill thousands of human beings at a single blow, and by extension many cities and millions of human beings, was entirely new, and required new ways of thinking and acting.

The reports from Hiroshima and Nagasaki of a strange, new malady that seemed to be affecting the survivors emphasized the point that nuclear weapons were unique. Manhattan Project scientists had known something of the dangers of the radiation that would be released by the Bomb; the fact that ionizing radiation had profound effects, mostly bad, on living organisms had been well-established biology by the 1930s. But they knew that the prompt radiation from the Bomb would be of little importance because most everyone who received enough radiation from an atomic explosion to kill them would already be killed by heat and blast anyway. Scientists knew that there would be residual radiation, especially from an explosion at ground level that contaminated soil and debris, and hurled it skyward to fall back to the ground at great distances, but in the rush to build the weapon, there was little time to contemplate or study such matters.

Now the results were showing themselves in real human beings. Initial Japanese reports of "an atomic bomb illness" were dismissed by Groves and other authorities as mere propaganda, but those who visited Hiroshima and Nagasaki after the bombings soon saw otherwise. Controversy would swirl around the issue for years afterward, but in any event, it was quite clear that here was a consequence

The ability to destroy an entire city and kill thousands of human beings at a single blow . . . was entirely new, and required new ways of thinking and acting.

of nuclear weapons that didn't exist with the conventional type.

The public too was unsure what to think. As the euphoria and relief of the end of the war faded, uncertainty and a certain amount of anxiety took their place. The press began to offer a steady stream of tales about the wonders that would soon arrive as the atom transformed from a weapon into a source of unlimited energy.

As political leaders, scientists, and soldiers all knew, though, it wasn't really possible to completely separate the civilian atom from its military counterpart. Reactors for peaceful electric power could also be used to manufacture fuel for weapons—an uncomfortable truth that complicated visionary ideas of sharing atomic technology with other nations. Again, it was apparent that some kind of controls would be needed on an international level to prevent a nuclear free-for-all. Somehow, policies had to be established for technical information on atomic matters, the distribution and administration of atomic materials, and deciding how much openness and security were needed in the atomic realm.

The first and most pressing issue was the USSR, which although gravely depleted and damaged by the war, was quickly emerging as the United States' immediate rival, not only politically and militarily, but in the atomic realm. Estimates varied regarding how soon a Soviet atomic bomb might be developed. Scientists such as Oppenheimer thought it might take as little as five years or less, while military leaders such as General Groves, more dismissive of Russian capabilities, said it would be ten or even twenty years before the Soviets could repeat the United States' achievement. Whatever estimate one chose to go with, few expected America's atomic monopoly to last forever. One way to preserve it, at least for a time, if not forever, might be to convince Stalin to renounce atomic weapons. But how to do that without doing so ourselves?

In early 1946, Truman appointed a committee to "consider the problems arising as to the control of atomic energy and other weapons of possible mass destruction," led by Undersecretary of State Dean Acheson and Tennessee Valley Authority chair David Lilienthal. Oppenheimer also served on the committee, beginning what would turn out to be a long string of postwar advisory appointments on various government agencies and committees that would cause trouble for him down the road. The Acheson-Lilienthal committee would eventually direct the results of its deliberations to the newly formed United Nations Atomic Energy Commission, consisting of all the members of the UN Security Council including the USSR.

Hopes ran high that somehow, a practical structure could be worked out to ward off the specter of unlimited atomic warfare. That specter had grown even more frightening as the full tragedies of Hiroshima and Nagasaki became more widely known. Groves and other US authorities had managed to prevent the dissemination of some of the more graphic images and accounts of the Bomb's effects, but enough got out to scare people. Their fears were exacerbated by articles in popular magazines detailing the effects of atomic bombs on US cities such as New York and Chicago. It was one thing to contemplate the devastation of unfamiliar, alien foreign cities, but quite another to ponder lurid images of familiar landmarks such as the Empire State Building consumed in nuclear fire.

International problems aside, there were also domestic issues to consider. During the war, the Manhattan Project had been under military control, and even with the war over, the military was loath to give up the vast empire that Groves had built. Military supporters in Congress pushed a piece of legislation called the May-Johnson bill, vigorously opposed by scientists because of its restrictive provisions on research and scientific openness. After much debate, alternative legislation was drafted, and after even more argument as well as backroom wheeling and dealing, it finally became the Atomic Energy Act, signed into law by Truman in summer 1946 and establishing the civilian-led Atomic Energy Commission (AEC).

That took care of atomic matters within the United States, but reaching an international arrangement was a much more complicated proposition. The Acheson-Lilienthal report that went to Truman at the end of March proposed an international agency to administer research

It was one thing to contemplate the devastation of . . . foreign cities, but quite another to ponder lurid images of familiar US landmarks consumed in nuclear fire.

and development while also possessing sole control over the world's stockpiles of nuclear materials such as U-235 and plutonium, making sure that they would be used only for peaceful purposes. The agency would have broad powers of inspection to ensure compliance too. After these proposals went into effect, the United States would give up its nuclear monopoly and share some (but not all) of its atomic knowledge with the world at large.

Following some modifications, and now called the Baruch Plan after the US emissary who presented it to the United Nations Atomic Energy Commission, the proposal enjoyed a generally positive response from most quarters save one: the Soviet Union, which didn't quite believe US assurances of giving up its atomic advantage to the United Nations. It didn't help that the United States was simultaneously preparing the first postwar atomic test series, Operation Crossroads on Bikini Atoll in the Pacific. If the United States was so sincere about peace and disarmament, asked Soviet representatives, why was it planning to explode more atomic bombs?

The Soviet Union soon countered with its own proposal for international atomic energy control, and the remainder of 1946 was spent in endless, tedious, sometimes harsh negotiation and argument. Meanwhile, the general tensions and distrust between the United States and USSR only increased, and by the end of the year, it was clear that the first major efforts at nuclear disarmament and the

Figure 3 The United States began Operation Crossroads, the first postwar atomic testing program, with the Able explosion in the Pacific on July 1, 1946. Note the ships moored around the detonation site.

international control of the atom had collapsed. It had been a noble quest, but perhaps too much too soon, and a brief opportunity that would never quite repeat itself.

Outside the political arena, the science and technology of nuclear weapons were only accelerating. Instead of being disbanded after the war as some had expected, the Los Alamos laboratory continued to operate, with some of its wartime staff remaining and dedicating themselves to improving and perfecting fission weapons. The idea of the Super or hydrogen weapon, though proving to be ever more challenging, hadn't gone away, having been kept alive by its tireless champion, Teller. The other Manhattan Project installations at Oak Ridge and Hanford kept going as well, now under the aegis of the new AEC. And an expanding program of atomic testing, beginning with Crossroads in 1946 and followed by Greenhouse in 1948, examined new bomb concepts and designs while also providing more data on the physics and effects of nuclear explosions.

Not that all this activity was completely unquestioned. As the United States began to gain some distance from Hiroshima and Nagasaki, and US society returned to normal after the wartime years, some started to challenge the official narrative that the atomic bomb had been necessary to end the war and prevent the US invasion of Japan. Hadn't Japan already been on the verge of surrendering anyway? If we had undertaken the atomic bomb project out of fear of Hitler getting it first, why did we continue after Germany surrendered? These and other troubling questions were encouraged by newly organized groups such as the Federation of Atomic Scientists (later the Federation of American Scientists), formed by a group of Manhattan Project researchers. Cutting their political teeth in the battles over US atomic energy legislation, such organizations provided a means for formerly apolitical scientists to make their voices heard regarding the far-reaching implications of the scientific weapons they had created. Other scientists such as Oppenheimer straddled the lines between the academic and political realms, attempting to advocate for scientific concerns while serving as high-level advisers to the government.

Most of the top scientists of the Manhattan Project had returned to their sedate academic lives after the war, satisfied that they had done their duty for their country. Oppenheimer was one of the exceptions. Although he gratefully relinquished his position as director of Los Alamos to return briefly to his old job at Berkeley before being named director of the prestigious Institute for Advanced Study in Princeton, he found himself increasingly involved in policy making, called to serve on various committees and advisory panels for the new AEC and other agencies. Atomic science was still a brand-new and mysterious world to most people in and out of government, and the unique expertise of men like Oppenheimer, Lawrence, Fermi, and others was indispensable to those struggling to formulate domestic and international policies, and answer the unprecedented questions that the existence of the atomic bomb now posed.

Any hopes that atomic research could now concentrate on peaceful applications rather than building bigger and better weapons were soon tempered by the ever-intensifying Cold War taking shape. As the USSR tightened its grip on the Eastern European nations that it had occupied since the end of the war and began trying to influence noncommitted nations to embrace Communism over Western capitalism, the United States abandoned any remaining ideas of preserving its wartime alliance with Stalin. Truman and his advisers declared a policy of "containment" to prevent the spread of Communism, with the implication that the atomic bomb would be its guarantor. Even when facing such blatant challenges as the 1947–1948 blockade of West Berlin that made it clear that Stalin was not about to play nice with the West, the United States and its allies could rest assured in the knowledge that even if the Soviets decided to invade Western Europe, the atomic bomb stood ready to stop them in their tracks.

Which made the question of when the Soviets might get the Bomb ever more pressing. The lack of any good intelligence sources that might provide some handy clues only increased anxiety. The newly formed CIA, created in 1947 from the remnants of the United States' wartime intelligence agencies, watched and waited, sifting through thin hints here and there that managed to somehow find their way out of the tightly closed society of the USSR.

Finally, on September 3, 1949, a specially equipped B-29 reconnaissance plane—a modified version of the same type of aircraft that had A-bombed Japan—detected the hard evidence that all had been dreading. As the plane flew near the Kamchatka Peninsula off of Siberia, filters collected radioactive dust particles borne on high-altitude winds blowing out of the Soviet Union. The particles were quickly analyzed and showed the telltale isotopic signature of an atomic explosion. US scientists analyzed the data and reported the inevitable conclusion: the USSR had the atomic bomb. It had been detonated around the end of August in a test that US officials dubbed "Joe 1," after Stalin.

The Hydrogen Age

Preempting any official Soviet boasting, President Truman announced the Russian bomb to the world on September 23. His terse comments downplayed the Soviet achievement, noting that it was nothing that the United States had not already anticipated. Behind the scenes, however, not to mention in public, some were already panicking at the supposed loss of US nuclear supremacy.

A few people tried in vain to deny or understate the news. Surely the Russians were too backward to have accomplished such a feat so soon after the United States. Perhaps, instead of a bomb test, we had only detected debris from a failed nuclear reactor or similar project. And even if the Soviets had a bomb, it obviously had to be inferior to ours. If it were true, though, then some kind of treachery had to be afoot. No doubt Soviet spies had stolen the secrets of the atomic bomb from the United States. (As we will see, this viewpoint, while not wholly accurate, was partially correct.)

The reactions of Oppenheimer and most of the other atomic scientists were considerably more restrained since they had been warning even before Trinity that the secret was not going to hold up for long. An exception was Teller, who hadn't been expecting a Russian bomb quite so soon. In considerable agitation, he called Oppenheimer as soon as the test was announced to ask him, "What should we do?" An annoyed Oppenheimer replied, "Keep your shirt on."[1]

It was not a conducive atmosphere for that, however. The United States immediately stepped up its production of atomic bombs to beef up its arsenal. Meanwhile, refusing to be mollified by Oppenheimer's low-key reaction, Teller soon realized that rather than a disaster, the Soviet bomb now presented him with the perfect opportunity to push his pet project. Research on the H-bomb had been

Figure 4 US Civil Defense poster. The Soviet acquisition of the atomic bomb increased American anxieties and brought renewed emphasis on civil defense. (National Archives and Records Administration)

continuing at a low level yet not making much progress. To Teller's immense dissatisfaction, no one seemed to feel much of a sense of urgency about the problem, not with the war won and the United States on top.

But for Teller and others within military and governmental circles, the Soviet bomb changed everything. It didn't matter that the United States undoubtedly still had more bombs than the Russians, and continued to build more and better ones; atomic equity on any level was completely unacceptable and an existential threat. The clear answer was to regain absolute superiority by building the hydrogen bomb.

Still, that conclusion wasn't quite so obvious to some. One opportunity to control and perhaps reverse the spread of atomic weapons to other nations in a dangerous spiraling arms race had already been lost with the failure of the UN efforts. Now here was another chance for the United States and world to forgo a terrible weapon that might lead to ultimate disaster. Even better, it was a chance to do so before such a weapon even existed. This wouldn't be the rest of the world renouncing something that only the United States had. It was a way to choose peace over possible thermonuclear war.

As had become the new norm when considering nuclear issues, the question of building the H-bomb was both scientific, political, and even moral, all inextricably mixed together in ways that couldn't be addressed simply. As heated debate raged within both the administration and public forums, again the government turned to its scientists—in this case, the General Advisory Committee (GAC) of the AEC, which was the AEC's chief scientific advisory body, chaired by Oppenheimer.

The GAC met at the end of October 1949 to consider the H-bomb question and issue recommendations to the AEC commissioners, who would then pass on their own recommendations to the president. Fermi, I. I. Rabi,

Conant, Seaborg, and several other distinguished experts, all immensely qualified to consider all the technical ramifications as well as the military, political, and ethical issues, joined Oppenheimer on the GAC. Over the Halloween weekend, the GAC considered, debated, and finally came to definite conclusions about how to respond to the Soviet atomic bomb. Some of their conclusions would arouse little argument, such as increasing the production of fissile material, accelerating research on "boosted" atomic weapons with greater yields, and developing more versatile and flexible military strategies.

But one recommendation was not going to be popular at all. Although not opposing further research on the hydrogen bomb, the GAC unanimously recommended against any kind of crash program to build it as soon as possible, as advocated by Teller and his colleagues. Their arguments against an H-bomb "Manhattan Project" were quite reasonable. It would divert resources and effort crucially needed to build and maintain the existing atomic stockpile to a project that might not even work; an H-bomb might not even be physically possible; and the United States already had a stockpile of hundreds of atomic bombs that were sufficient to meet any conceivable threat for the foreseeable future. And assuming that the H-bomb could be built, what would be its purpose? It would be so powerful and destructive that it could only be used on cities as more of a genocidal device than a practical military weapon. Several of the GAC members called the H-bomb "an evil thing considered in any light."[2]

The final decision, of course, rested with the president, who had to weigh moral consideration with practical geopolitical realities. Truman also faced enormous pressure from the military as well as hawkish influences within his own administration, such as his secretary of state, Acheson. And Communist influence only continued to grow and spread across the globe. Aside from an increasingly belligerent Soviet Union, the United States now met with a newly Communist China, forming a seemingly united and implacable "Red Menace" opposed to freedom and bent on dominating the world. As if to make matters even worse, Communist spies were suddenly appearing everywhere, with disturbing connections to the atomic bomb. In England, Klaus Fuchs, a German-born physicist who had worked as part of the British contingent of Manhattan Project scientists at Los Alamos, confessed to spying for the Soviet Union, while another extensive spy ring was uncovered in Canada. The free world, it seemed, was under siege from all sides.

In such a climate, it wasn't surprising that on January 31, 1950, Truman announced, "I have directed the Atomic Energy Commission to continue its work on all forms of atomic weapons, including the so-called hydrogen or superbomb."[3] Although Truman was only affirming official support for work that was already ongoing, the

Joint Chiefs of Staff and H-bomb advocates in the administration and scientific community soon made sure that the president's directive was transformed into an all-out crash effort to build the H-bomb. Rather than resist such attempts, Truman supported them. When the Cold War between East and West finally went hot with the invasion of South Korea by Soviet-backed North Korea in June 1950, the urgency of maintaining atomic supremacy by achieving the H-bomb seemed even more necessary.

The problem was that despite the enthusiasm of Teller and his compatriots, no one actually knew how to do it. Intricate calculations and models of the characteristics of thermonuclear explosions, now possible thanks to newly developed electronic computers with names such as MANIAC, ENIAC, and UNIVAC, soon demonstrated that the initial concept of the H-bomb developed at Los Alamos during and after the war, known as the "classical Super," was unworkable.

After Truman's decision made the project official, some of the physicists who had strongly opposed the H-bomb had decided to work on it anyway in hopes of perhaps proving once and for all that it was impossible. For a time, it seemed that they were getting their wish. The hydrogen bomb appeared to be nothing more than a fascinating scientific idea that would remain nothing further.

Another series of atomic tests in the Pacific in spring 1951 included some experiments that provided more data on thermonuclear reactions, but still didn't offer the needed breakthrough required for a practical weapon. Later in the year, however, Stanislaw Ulam, a Polish mathematician and Manhattan Project veteran now working on the H-bomb effort, came up with a new design that seemed to be the answer. Teller and Ulam further refined the idea into what became known as the Ulam-Teller (or Teller-Ulam, depending on who you talked to) configuration.

More calculations and practical design work followed, until by late 1952, the concept was ready to be tested. It wasn't quite yet an operational weapon. The liquid hydrogen fuel for the device required a cryogenic plant to keep it cold, and the entire sixty-five-ton apparatus filled a large warehouse-like structure on the tiny island of Elugelab within Eniwetok Atoll. But it would be enough to prove the Ulam-Teller configuration. The test was named Mike, the first shot of Operation Ivy.

Shortly before dawn on November 1, 1952, the firing signals were sent and the device was detonated in the world's first thermonuclear explosion. The small island was vaporized by slightly over ten megatons of energy, with a fireball three miles in diameter and a mushroom cloud that reached an altitude of 118,000 feet. The United States had its hydrogen bomb. It was back on top.

Unfortunately, that comforting sense of superiority would not last long. Less than a year later, on August 12, 1953, the Soviet Union detonated its own thermonuclear weapon. Again, there was a hue and cry over espionage with the conviction of Fuchs, and now the trial of electrical engineer Julius Rosenberg and his wife, Ethel Rosenberg, both members of the Communist Party of the United States of America. Surely, some claimed, all this proved that the Russians had cheated. But just as the USSR had already been working on its own atomic bomb before Hiroshima, their gifted scientists, chief among them Igor Kurchatov and Andrei Sakharov, had also been quite aware of the hydrogen bomb concept and working actively on it for quite some time. Certainly the expert espionage campaigns of the USSR had helped Kurchatov and his colleagues around some of the difficult spots, and provided valuable clues and shortcuts, but the common notions at the time that the Russians had simply "stolen" and copied the United States' bombs were complete misconceptions.

If the US public wasn't aware of such facts as yet (and with certain parties within the government and military perfectly content to let them imagine the worst), the scientists who examined data from Soviet tests, mostly in the form of radioactive debris collected from sampling aircraft, knew otherwise. Although the August 1953 test had not been a full-fledged hydrogen bomb but instead only one that utilized thermonuclear principles, with a yield much smaller than Mike, it was actually a more sophisticated design not requiring an awkward cryogenic fuel. The Soviets might have been behind the United States, but they had figured out a few tricks of their own too.

Not that such fine distinctions mattered to most people. Now both the United States and USSR had the hydrogen bomb—a weapon that could be made as powerful as desired, and could wipe out not merely a city but also an entire urban area at one blow. That job was up to the US Air Force since its founding in 1947, and specifically its Strategic Air Command (SAC). New and versatile bombers were developed specifically to handle the atomic bombing mission, with the B-29s that had devastated Japan soon replaced by the massive B-36, and later, jet-powered aircraft such as the B-47 and B-52.

With the Soviets acquiring their own nuclear weapons and America's primacy long gone, the floodgates of unlimited testing were unlocked. The first public test program, Crossroads in 1946, featured only two atomic blasts, and its 1948 follow-up, Sandstone, only three. All were conducted far from the United States on Pacific islands. In January 1951, however, with Los Alamos again humming with activity on new bomb designs and the H-bomb project, the testing program expanded to include not only the Pacific sites but also the continental United States with the establishment of the Nevada Test Site, about sixty-five miles northwest of Las Vegas. On January 27, Able, the first atomic explosion on US soil since Trinity, was conducted, dropped from a B-50 bomber and lighting up the

Nevada dawn with a one-kiloton burst. Four more shots would follow over the next couple of weeks.

It was the opening salvo of a series of continental atomic tests, mostly in Nevada, that would continue for over a decade, testing weapons designs, effects, scientific principles, and the psychological reactions of US troops exposed to nuclear detonations at close range. Although the Nevada tests were restricted to relatively low yields—the largest Nevada test in July 1957 had a yield of seventy-four kilotons—fallout from the one hundred tests spread across nearly all the United States throughout the years of testing.

The big weapons, including the H-bombs in the megaton range, were reserved for the more remote Pacific test areas, not only because of their much higher yields, but because of their far more extensive fallout. It was one thing to dump radioactive debris on unwitting US citizens in Poughkeepsie or Chicago, yet no one much cared if it fell onto (mostly) uninhabited islands or into the open ocean.

Fallout was one of the phenomena that made nuclear weapons more frightening to the average person than conventional bombs. The idea of radiation—something that can't be seen or felt, but that could be dangerous and even fatal—was a particular concern to the public, and the AEC and military took painstaking steps to assuage anxieties over it. Such efforts were more or less successful in the early years of the atomic age, but as ever more powerful weapons were tested with increasing frequency, it was inevitable that official reassurances were going to directly conflict with physical realities. That happened with the test of the United States' first practical hydrogen bomb in the Bravo test of Operation Castle on Bikini Atoll on March 1, 1954.

Bravo surprised everyone, beginning with its designers who had been expecting a modest thermonuclear yield. Instead of the five megatons predicted with this new bomb design, Bravo was a fifteen-megaton monster, its fireball expanding so rapidly and quickly that some observers feared it might not stop at all. Adding to that nasty surprise were shifting wind patterns that sent the unexpectedly huge cloud of fallout over vast unprepared areas, including one where a civilian Japanese fishing vessel, the *Lucky Dragon*, had been working. The twenty-three crew members were doused with debris from Bravo and soon became ill with radiation sickness. One of them later died after the vessel returned to port. It was an unprecedented public relations disaster for the AEC and United States, leading to media outcry along with diplomatic and public protests that were impossible to ignore.

Although the initial furor eventually died down, the Castle-Bravo incident proved to be the impetus for a worldwide grassroots campaign against nuclear testing specifically and nuclear weapons in general. In 1955, British philosopher Bertrand Russell and Einstein released a joint statement, the Russell-Einstein Manifesto, warning of the dangers of hydrogen warfare and calling for an international scientific conference to discuss possible solutions. It turned out to be Einstein's last public statement before his death, and led to the Pugwash conferences on disarmament and global security, which brought together scientists from East and West, and eventually laid important groundwork for later international agreements.

Other public protests had more immediate and less exalted concerns. The fear of a thermonuclear war between the United States and USSR that might end the world was bad enough, but as scary as it was, it was only a possibility. Radioactive fallout from atomic and hydrogen bombs being routinely exploded in the open air, falling onto our cities and homes, and working its way into our food, milk, and the bodies of our children was happening now. And the authorities didn't seem to care, content with issuing hollow reassurances, and telling the public that building and testing bombs was the only way to stay safe as well as preserve the peace.

As if to emphasize the futility of any ideas of control, the Bomb began to spread beyond the bipolar influence of the United States and USSR. Britain became the third nuclear power in October 1952 with its first atomic bomb, following it up with its own hydrogen bomb in May 1957. Not to be left out of the growing club of nuclear powers, France was working on its own bomb, finally achieving it in 1960. All along, the United States and USSR—now universally acknowledged as "superpowers" with the advent of the hydrogen bomb—continued to steadily increase their own arsenals, tirelessly refining the weapons, on the one hand, while devising and perfecting ever more flexible ways to use them, on the other.

The atomic age had started with fear and apprehension, but in the beginning at least, shining through all the talk of doomsday, there had been a thread of hopefulness and optimism, a feeling that it might still be possible to transform this terrible weapon into a force for good that could unite the world and make war impossible for all time. Those dreams now seemed naive and unattainable. Even though, bowing to public and political pressure, the United States and USSR agreed to an informal moratorium on nuclear testing in 1958 as both sides attempted to negotiate a formal test ban treaty, the USSR eventually broke the moratorium in 1961 and testing resumed.

So the Bomb was here, and it would remain. As the 1950s passed into the 1960s, the world seemed to be on an irrevocable and ever-accelerating course toward self-destruction. More than once, the always simmering tensions between the United States and USSR threatened to boil over into the apocalyptic war that everyone feared but no one really wanted.

THE BRINK: CLOSE CALLS AND NUCLEAR CRISES

By the mid-1950s, the Bomb had shifted in the public consciousness from a novel phenomenon into an ever-present part of everyday life. Talk of kilotons and megatons, fallout, and hydrogen warfare could be found in popular newspapers and magazines, ranging from news reports on the latest test shots in the Pacific or Nevada, terse speculations on what might happen if war came, reports on the most recent Soviet provocations in Eastern Europe or Asia, and learned commentary on the United States' strategic options and military plans.

Running throughout it all was a steady current of anti-Communist rhetoric that included affirmations of the US way of life, condemnations of the godless Reds, and accounts of the latest accused Communist spies and subversives uncovered in the government or elsewhere in the public sphere. One of the most prominent of these accounts was Robert Oppenheimer himself, barred from any further governmental role in 1954 when political and personal enemies, among them his former Los Alamos colleague Edward Teller, maneuvered to have him permanently stripped of his security clearance.

The sharp lines between East and West, Communism and capitalism, totalitarianism and freedom, had become sharply defined. It was quite clearly us versus them, good versus evil. Churchill may have proclaimed the descent of an "Iron Curtain" over Europe, and by extension across the rest of the world, in a 1946 speech, but what had at that time been mostly a dramatic rhetorical device was now reality. And there was no greater symbol of that reality than nuclear weapons.

Before the advent of the hydrogen bomb, when it was still possible for some people to think of the atomic bomb as just another (if slightly more exotic) weapon, and particularly before the Soviet Union got it, there had been talk of using atomic weapons. During the height of the Berlin blockade of 1947–1948, some of the specially modified "Silverplate" B-29s, which were known as the only aircraft then capable of delivering atomic weapons, were deployed to Britain as a veiled warning to Stalin of what might be in store should he continue his intransigence or escalate to a military confrontation. It was only an empty gesture since the airplanes weren't carrying bombs, nor did the United States even possess more than a few operational bombs at the time (an uncomfortable truth that of course was kept under the strictest secrecy). When Communist Chinese troops entered the Korean War with a devastating counterattack on UN forces at the end of 1950, President Truman publicly hinted at a possible use of atomic weapons to strike back. Inheriting the Korean conflict after his 1952 election, President Dwight D. Eisenhower also considered using "tactical" nuclear weapons against the Chinese to break the bitter military stalemate that had set in as the war dragged on.

Such proposals never moved beyond the stage of hypothetical planning, but they were a definite sign that the idea of a taboo against the use of nuclear weapons had not quite set in. The hydrogen bomb, however, changed the game; its sheer power and "genocidal" quality, as the GAC had termed it, put it in a wholly different class. An atomic war between the United States and USSR, as awful as it would have been, might have been conceivable in some sense as a war that could be won or lost because each nation might still survive in a recognizable, viable form. Yet if H-bombs began dropping from the skies, all bets were off; neither country would be the victor, and the entire society of each would be destroyed by any practical definitions. Thermonuclear warfare was all or nothing.

With East and West now settled into a Manichaean confrontation, continually adding to their arsenals and challenging each other in every fundamental area of politics, military posture, and even culture, it was inevitable that this uneasy, tenuous stability would occasionally be tested in crisis. Some tests would be minor, almost over before they started. Some would remain buried in secrecy for decades. And others would be known and experienced by the entire world as they happened. Some were resolved through adept diplomacy and conscientious action, but many avoided tipping over into disaster only through sheer luck. Almost all of them could have been prevented from ever happening in the first place, but for human shortsightedness, fear, and suspicion.

The Shadow of Armageddon

Aside from being the first time that East and West had directly clashed militarily in the Cold War, the Korean War was the first instance in which the United States had to face the reality that it could no longer decide to use nuclear weapons with impunity. Although Eisenhower once remarked that in situations where nuclear weapons could be used against strictly military targets, "I see no reason why they shouldn't be used just exactly as you would use a bullet or anything else," he also knew that the United States had to take into account the probable reaction of the USSR: either direct nuclear retaliation on US targets or more likely a Soviet invasion of Western Europe.[1]

With East and West now settled into a Manichaean confrontation, . . . it was inevitable that this uneasy, tenuous stability would occasionally be tested in crisis.

Whichever happened, it meant likely escalation to a full-scale world war between the United States and USSR. Any short-term military gains were meaningless when weighed against that possibility. The same considerations held for later East-West confrontations throughout the 1950s over Taiwan and in the Middle East.

Aside from public confrontations, the United States and USSR engaged in hundreds of top secret military skirmishes throughout the 1950s and early 1960s. Facing the impenetrable enigma of the Soviet Union, military strategists realized that there was only one way to obtain vital intelligence on Russian military capabilities. Long-range bomber aircraft, the same planes that might deliver nuclear doomsday, were stripped down, fitted with sophisticated cameras and electronic surveillance equipment, and sent to patrol along the periphery of the USSR in international airspace, photographing whatever the cameras could see, sniffing out and recording radar signals and classified communications, and occasionally even venturing briefly inside the border to rouse Soviet fighters.

The Russians conducted similar excursions along US borders and overseas bases. The unarmed spy planes were routinely intercepted and challenged by fighter planes, and often fired on—and more than once, actually shot down over Soviet territory. A sort of gentleman's agreement ensured that these completely illegal and unauthorized incursions were kept strictly secret by both sides to avoid international political repercussions. In the far-flung locales where such dramas took place, the occasional loss of an aircrew could be explained when necessary as an unfortunate accident. But a major incident finally occurred on May 1, 1960, when pilot Francis Gary Powers's U-2 spy plane was shot down deep inside the USSR, scuttling an important East-West summit meeting in Paris.

By the late 1950s, another source of constant tension and possible disaster emerged: a false attack warning that triggered a nuclear response. For the first decade or so of the Cold War, even as strategic bombers evolved from propeller driven to jet powered, and increased their range and speed, they needed hours to reach their targets. That allowed at least some time for contemplating, considering, and even aborting an attack. There would be some chance for leaders to negotiate.

And although the nightmare of US strategists was an atomic Pearl Harbor, called a "bolt out of the blue" attack, enemy aircraft could still be detected by the Distant Early Warning Line of radar stations stretching across North America from the upper United States to the Arctic, so that the dedicated airmen of the Tactical Air Command, SAC's poorer sibling within the air force, could go into action and shoot down some, if not all, of the attacking force before our cities were obliterated.

But this relatively comfortable status quo began to change rapidly when the Soviets tested their first ICBM in

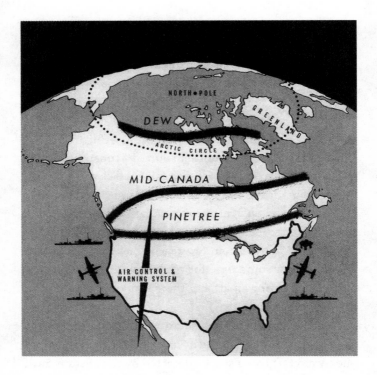

Figure 5 An elaborate network of early warning radar stations stretched across North America to warn against surprise attack, but also increased the possibility of false alarms. (Wikipedia EN Portal)

summer 1957 and then proved its ability by launching the world's first satellite, Sputnik, that October. Their second satellite a month later was far larger and heavier (carrying a doomed canine passenger named Laika). To the general public, such feats may have injured US prestige, yet for US strategic planners, the more ominous lesson was that the USSR now had the capability to hurtle hydrogen warheads down on North America. Although the United States had been busily developing its own missiles, Russia was far ahead. ICBMs were not only unstoppable but they could also reach their targets anywhere in the world in less than an hour. The short though measurable window of warning that the United States had enjoyed was gone.

Frantic reports and studies within the Pentagon and military think tanks warned that the United States was now open and vulnerable, and the danger would only intensify over the coming years. SAC rushed its ICBMs into production and deployment while continuing to build up its fleet of bombers, operating on what was essentially a war footing. SAC commander General Curtis LeMay, who had previously masterminded the bombing campaign against Japan including the atomic strikes, emphasized to his men that "we are at war now"—no matter that no one was actually dropping any bombs yet.[2]

Now armed with more nuclear weapons than ever before, SAC and its Soviet counterparts were poised on constant alert. With warning times cut from hours to

Figure 6   General Curtis LeMay, the pugnacious commander of the Strategic Air Command, kept his forces on a constant near-war footing. (National Archives)

minutes, the chance for a fatal error that might spark a war increased exponentially. Both sides acknowledged that they would now be forced to respond almost instantaneously to any threatening move by the other, with only the smallest margin remaining to confirm orders to retaliate. As a by-product of such tension came the specter of being preempted: having your own forces wiped out before they could strike back under attack. To counter that possibility, SAC kept many of its bombers and missiles at their highest state of readiness, endlessly drilling to hone to a fine edge their ability to be launched in minutes. In the early 1960s, in a program code-named Chrome Dome, SAC instituted a policy of keeping a minimum number of nuclear-armed bombers in the air twenty-four hours a day, always ready for action.

These considerations also encouraged both sides to devise new ways to preserve their capacity to retaliate. For the Soviet Union, a nation that had been traumatized by the devastating Nazi invasion of 1941, this was a particularly sensitive problem. Unlike the United States, the USSR didn't have conveniently situated allies on whose territory to place its aircraft and missiles, limiting its ability to decisively strike the continental United States. Its decision to finally correct that strategic deficiency with a risky and dangerous gamble led to the greatest crisis of the Cold War—one that brought the world within hours of World War III.

Armageddon Ninety Miles Away

When Cuban revolutionary Fidel Castro first came to power in 1959, he kept the world guessing for awhile. At first Castro seemed committed to holding free elections and establishing a democratic state, but by the time that Kennedy became president in 1961, Castro had set up his own oppressive government and declared Cuba a socialist ally of the Soviet Union.

Efforts to oust Castro through both covert CIA actions and military invasion by US-backed Cuban refugees at the Bay of Pigs in April 1961 failed, and Cuba remained an open sore for the United States—a Communist power only ninety miles from the country's shores. Soviet leader Khrushchev soon took advantage of the opportunity to establish a foothold in the Western Hemisphere, embracing Castro's regime, and providing economic and military aid, much to US consternation. In spring 1962, following a summit meeting with Kennedy the previous year in which he had concluded that the young president was weak and easily intimidated, Khrushchev took his pledge to defend his Cuban ally to a new level. He decided that he would secretly deploy nuclear missiles on the island.

Not only would such a move help to ensure Cuba's security, but it would give the USSR a quantum leap in strategic advantage since the long-range ICBMs based in the Soviet Union were still small in number, and not yet very reliable or accurate. It was a risky move to be sure, but Khrushchev gambled that he would be able to pull it off directly under the nose of the United States.

A massive effort ensued throughout summer 1962 with the USSR moving missiles, people, and equipment to Cuba disguised as routine cargo. Kennedy publicly warned Khrushchev that "the gravest issues would arise" if the

USSR placed offensive weapons in Cuba, while Khrushchev insisted the shipments were merely defensive, intended to deter any further US threats. US surveillance aircraft noted the deployment of antiaircraft surface-to-air missiles, but no definite evidence of anything more serious.

On October 14, 1962—ten days after the first nuclear warheads arrived—a U-2 spy plane flew over western Cuba and photographed launch sites under construction near the town of San Cristobal. Also clearly visible in the more than nine hundred reconnaissance photos were launchers, support equipment, and SS-4 medium-range ballistic missiles, known to have a range of up to approximately a thousand miles—enough to put Washington, DC, within easy range. A few days later, SS-5 intermediate-range missiles with a range of about 2,200 miles were also detected.

Informed by US national security adviser McGeorge Bundy, Kennedy wasted no time calling together a team of top-level advisers, dubbed the Executive Committee or Ex Comm, to deliberate how to respond. The first reaction was a combination of astonishment and anger—the former over Khrushchev's gall to take such a risk, and the latter over his obvious deception and repeated lies.

The United States had caught one break. As CIA experts affirmed, the missiles were not yet operational and ready to fire. The United States had discovered the Soviet deception before any fait accompli was achieved. But work on the missile sites was continuing around the clock, meaning that the window of opportunity to take action was closing rapidly. Kennedy and the Ex Comm considered, debated, and argued over the options.

The most impulsive and perhaps satisfying initial option was a military one: an immediate attack on the launch sites to wipe them out before they were ready. But that would be a more complicated, messier solution than it seemed on first glance. Air strikes couldn't ensure the complete destruction of all the sites and would have to be followed with an invasion by ground forces. Moreover, any military action undoubtedly meant killing not only Cubans but Russian nationals too. How would Khrushchev react if the United States killed hundreds of his soldiers and technicians without warning?

Diplomacy was another possibility. The problem, as some of Kennedy's advisers pointed out, was that the Soviets could waste time pretending to negotiate while work on the missile sites went on. And once they were finished, why would Khrushchev give up such an advantage?

A third option emerged that offered a sort of compromise between hard military action and soft diplomacy. A naval blockade could prevent further missiles and warheads from reaching Cuba. It was an imperfect solution that did nothing to remove the missiles already there and was technically an act of war, but it was less drastic than an open attack and more firm than begging to negotiate.

After further debate and argument, Kennedy decided on the blockade (describing it with the softer, more peaceful-sounding term "quarantine"), reasoning that if it didn't work, the other options remained on the table. Meanwhile, US strategic forces were brought to increased alert while preparations began for air strikes and a possible invasion. One way or another, Kennedy resolved, Soviet missiles were not going to remain based in Cuba.

Raising the Stakes

For the first week after the discovery of the missiles, the situation had remained unknown outside the councils of power. That changed on Monday evening, October 22, when Kennedy addressed the nation and world on live television. Unmistakable evidence had established the fact that offensive missile sites were in preparation on Cuba, he announced, with the purpose of providing a nuclear strike capability against the Western Hemisphere. In response, the United States was imposing a quarantine on offensive military equipment to Cuba, placing its forces on increased alert, and calling for diplomatic action through the United Nations and Organization of American States, among other significant moves. Most important, Kennedy warned, "it shall be the policy of this nation to regard any nuclear missile launched from Cuba against any nation in the Western Hemisphere as an attack by the Soviet Union on the United States, requiring a full retaliatory response upon the Soviet Union."[3]

The lines had been drawn and the stakes unequivocally defined, not merely for the United States and USSR, but for the entire world. The Cold War, which for most Americans had been mainly only a constant yet dim background noise in the more immediate mundane concerns of their daily lives, had suddenly become a real and present danger. All across the country, people stockpiled groceries, stripping supermarkets of food and essentials, preparing for the worst. In other Western countries such as France and Great Britain, bewildered citizens staged demonstrations against both the United States and USSR for threatening world peace with their needless posturing, while in the USSR, the state-controlled media meant that few, if any, Soviet citizens knew that a crisis even existed.

As the US Navy deployed its quarantine around Cuba and Soviet supply ships inexorably continued to approach, the stage for direct military confrontation was set. US forces around the world had been placed on DEFCON (DEFense CONdition) 3 with President Kennedy's October 22 address, but two days later, that alert was raised to DEFCON-2—one step away from DEFCON-1, or actual war.

That was the precipice on which both nations balanced on Wednesday, October 24, as Soviet vessels encountered US Navy warships in the waters of the Caribbean. Russian warships were also present in the form of submarines, which—unknown to the United States until many years later—were armed with nuclear-tipped torpedoes.

For a few tense hours that day, it seemed that the first shots of World War III would be fired at sea, but somehow both sides pulled back from open confrontation. Russian vessels either turned around or submitted to inspection, and the first test of the Cuban missile crisis passed without incident. In the words of Ex Comm member US secretary of state Dean Rusk, both sides were "eyeball to eyeball, and the other fellow just blinked."[4]

That sense of relief didn't last long, however. More Russian ships were approaching the quarantine line, work on the missile sites in Cuba showed no signs of slowing down, and Khrushchev remained as defiant as ever. Meanwhile, much of the world, even including the United States' allies, remained unconvinced of the US allegations and presence of the missiles. Kennedy and his military advisers were reluctant to publicly release top secret U-2 reconnaissance photographs, but realized that keeping hard evidence under wraps also allowed the Soviets to continue denying the existence of the Cuban missiles while accusing the United States of provoking the crisis. What price secrecy when the fate of the world might be at stake?

At the United Nations on October 25, US ambassador Adlai Stevenson surprised the Soviets by openly displaying a series of the U-2 photos, revealing to the world the

For a few tense hours that day, it seemed that the first shots of World War III would be fired.

"unmistakable evidence" that Kennedy had decried earlier in the week. No one, whatever their political persuasions, could now pretend that the matter was only an extreme case of East-West posturing. It was a real threat.

By the time the week ended, the crisis had bifurcated into two distinct threads, one of them public, and the other secret. The public face, well represented (at least in the West) on the front pages of newspapers and frequent TV and radio broadcasts, consisted of the ongoing dramatics of the naval quarantine, and the details of which ships were stopped and which were allowed to sail on to their Cuban destinations. There were also the ongoing diplomatic wranglings in the United Nations; street-level demonstrations around the world in support of the United States or USSR, or opposing both; and the helpless anxieties of millions following the latest developments, hastily putting together makeshift fallout shelters in their basements, and wondering if the world was about to end.

The other thread of the crisis, rapidly becoming more and more dominant, was the secret correspondence between Kennedy and Khrushchev. Communications between world leaders were still slow and inefficient in the early 1960s, relying mostly on cable messages that had to be carefully composed, encrypted, transmitted, and then laboriously translated on the other side—a process that took hours at best, and a day or two at most. No "hotline" for instantaneous exchange between the US president and

USSR chairman yet existed. In one way, this was an advantage since it allowed more deliberate and thoughtful communication in absolute confidence between Kennedy and Khrushchev, with both free to speak their minds without worrying about political second-guessing from outsiders. But in a crisis such as this one in which matters continued to accelerate and threatened to spiral out of control, it allowed more time for the misinterpretation of motives, and less time to correct mistakes and missteps.

On Friday, October 26, Khrushchev dispatched a long message to Kennedy. The meandering personal letter mused on the costs of war and Khrushchev's sincere desire to find a way out of the present situation through "statesmanlike wisdom." He proposed a possible solution: he would remove the Cuban missiles in exchange for the removal of the naval quarantine and a US pledge to never invade Cuba. At least on the surface, it seemed to offer the first real possibility out of the impasse.

And then, before Kennedy and the Ex Comm had finished digesting and debating this first substantial communication from Khrushchev, a second arrived less than twenty-four hours later, on what would come to be called "Black Saturday." More terse and stern in tone than the first communiqué, this public communiqué added the condition that the United States remove its "analogous" nuclear weapons from the soil of its NATO ally Turkey before the USSR would withdraw its Cuban bases. Though annoyed at the additional demand, Kennedy acknowledged that "most people would regard this as not an unreasonable proposal. . . . [P]eople will think this is a rather even trade."[5] He knew too, though, that the public appearance of seeming to desert an ally under Soviet pressure would be politically disastrous.

Then two more immediate incidents quickly pushed aside thoughts of political complications. Thousands of miles away from Washington or Moscow, a US U-2 on an air-sampling mission accidentally strayed across the Arctic frontier of the USSR. US fighters managed to find the U-2 before Soviet fighters could intercept and guided its lost pilot safely home to Alaska. The tense incident highlighted the unfortunate fact that it was impossible to control every move of large military forces spread all over the globe. As Kennedy ruefully remarked on hearing of the incident, "There's always some son of a bitch that doesn't get the word."[6]

The second mishap of that Saturday didn't end so innocuously. It also involved a U-2, but one that was intentionally conducting reconnaissance over hostile ground. Flying over Cuba to collect additional photos of the still-under-construction missile sites, US Air Force major Rudolf Anderson Jr. was shot down and killed by a Soviet surface-to-air missile. When the report hit the White House, the pressure on Kennedy and his advisers instantly ratcheted up to an almost unbearable level. Blood had been spilled, and the Soviets had unquestionably fired first. Some kind of unequivocal military response, such as at least the immediate destruction of the surface-to-air missiles site responsible, seemed unavoidable.

War now seemed closer than at any other point since the crisis began. US forces were amassing in Florida preparing for a full-scale invasion of Cuba within a matter of days. SAC bombers were stationed at their prearranged points around the borders of the USSR, fully armed with H-bombs and awaiting only the order to proceed. Soviet submarines squared off against US Navy warships in the Caribbean. Open diplomatic efforts in the United Nations had stalled, while the confidential exchanges between Kennedy and Khrushchev seemed to have broken down in confusion and contradiction.

No one wanted war, but it seemed to be rolling inexorably from some dark, nameless place to engulf the planet in an unstoppable tsunami. After spending most of the day conferring with his president and the Ex Comm, debating responses to Khrushchev, and making final preparations for a military assault on Cuba, Secretary of Defense Robert McNamara recalled leaving the White House late that Black Saturday night and wondering whether he would live to see the sun come up the next morning.[7]

Yet the sun did rise again the next morning, and brought with it something that no one had been expecting. Early that Sunday came a new message from Khrushchev,

Figure 7 President Kennedy (*left*) confers with the Ex Comm on October 29, 1962, including Secretary of State Dean Rusk (*center*) and Secretary of Defense Robert McNamara (*bottom right*). (National Archives)

not via another cable, but in a public radio broadcast. He announced that he had ordered the dismantling of the Cuban missile bases along with the return of all missiles and warheads to the Soviet Union.

Khrushchev's reversal wasn't the complete capitulation it seemed to be at first to some parties. In a confidential response to the two letters that the Soviet premier had sent over the past couple of days, Kennedy agreed to a pledge to not invade Cuba and secretly remove US missiles from Turkey some months later. Kennedy's proposals had been delivered partly in a written response to Khrushchev, and partly through back-channel, one-on-one meetings between JFK's brother, attorney general and Ex Comm member Robert F. Kennedy, and the Soviet ambassador Anatoly Dobrynin. The Kennedys stressed that the Turkish part of the agreement had to be kept secret to avoid any appearance that the United States had "sold out" its NATO ally.

Technically, the crisis was not quite over; there were various diplomatic and logistical details to hammer out. The Strategic Air Command remained at DEFCON-2 status, and the naval quarantine continued until the end of November. But for the world at large, the Sunday rapprochement meant that everyone could relax, take a deep breath, and abandon their fears of impending Armageddon.

Lessons and Lucky Breaks

The nerve-racking "Thirteen Days" of the Cuban missile crisis, as they were later dubbed in a memoir by Robert Kennedy, generated ripple effects large and small, some of which persist into the twenty-first century. The most immediate result of the crisis was a distinct lessening of Cold War tensions, at least for a time. Both sides had crept all the way to the very edge of eternity, peered over, and abruptly withdrew from the dark oblivion they both saw clearly.

Realizing the critical need for faster and more reliable communications in case a similar crisis ever again arose, the United States and USSR established a dedicated hotline between Washington and Moscow, consisting of a dedicated telegraph/teleprinter line in operation twenty-four hours a day. Seeking a more substantial accomplishment to reduce tensions and build some measure of trust while addressing public concerns of nuclear fallout that had grown over the past decade, Kennedy and Khrushchev also signed the Limited Test Ban Treaty in 1963, banning all nuclear tests in the atmosphere, underwater, or in space.[8]

Those positives, while significant, were offset by negatives. Ignoring the fact that Khrushchev, despite all his blustery provocations throughout his tenure, had skillfully navigated not just the Cuban crisis but also various other confrontations and somehow managed to avoid World War III, hard-liners within the Soviet government chastised him for "weakness" and eventually forced him out of office in 1964. Relations between East and West under Khrushchev had been stormy and never simple, but he had shown admirable restraint, understanding, and in his own words, a "statesmanlike wisdom" in his dealings with the United States, particularly in the Cuban crisis. Now the United States would be forced to face hidebound Communist hard-liners, moving ongoing superpower relations into an uncertain realm.

That led directly to one of the most significant consequences of the missile crisis: a Soviet resolve to never again find itself in such an inferior position. As one Soviet diplomat remarked to a US counterpart, "You Americans will never be able to do this to us again."[9] Whereas the lesson that Kennedy and Khrushchev drew from the crisis was the need to reduce tensions and work toward disarmament, the new Communist leaders came to the opposite conclusion. The USSR had ultimately been forced to back down and remove its missiles because the United States still held unquestionable military superiority, meaning more nuclear weapons, more bombers and missiles, and more ways to deliver them with greater flexibility and accuracy. So the only possible response was an all-out effort to achieve strategic equity and finally superiority over the United States.

It would take some time for the ramifications of that response to reveal themselves as the Soviets developed and began fielding new weapons systems throughout the 1960s and 1970s. Meanwhile, as new information about those thirteen days in 1962 became known through both declassification and scholarly investigation in the following years, the chilling realities of the crisis came into ever-sharper focus. In the immediate aftermath of October 1962, Kennedy cautioned his fellow Ex Comm members to avoid any talk of "winning," and forgo acting or thinking of themselves as "victors." He considered it arrogant to think that the United

States had somehow won this confrontation with the Soviet Union because it was better and smarter somehow. The world had avoided nuclear war in this instance largely because of sheer luck rather than the brilliance of its leaders. Secretary of Defense McNamara confirmed that judgment decades afterward after speaking with his Soviet counterparts, reflecting in the Errol Morris documentary *The Fog of War*, "It was luck that prevented nuclear war. We came that close to nuclear war at the end [*gestures by bringing thumb and forefinger together until they almost touch*]. . . . Rational individuals came that close to the total destruction of their societies. And that danger exists today."

One can argue that of course Kennedy and Khrushchev were perfectly aware of the consequences if they had allowed the crisis to proceed all the way into thermonuclear warfare, and since no reasonable human being would choose that path, the events of October 1962 were not really the "crisis" they seemed to be. But perspectives of actual participants such as McNamara belie such tidy conclusions. In a situation where two heavily armed participants are poised on such an edge, neither completely sure of the other's intentions yet keenly aware of the threat they pose, rationality is not always the deciding factor, especially when the stakes are so high, and the response times so brief and automatic.

I've examined several moments in which the Cuban crisis might have sparked into open conflict: the first in-

The world had avoided nuclear war in this instance largely because of sheer luck.

terceptions of Russian ships, the accidental U-2 overflight of the Soviet border, and the shooting down of the U-2 over Cuba. As historians and political scientists have revealed, however, a number of other incidents occurred during those two weeks, most of them relatively minor, but ones that might have taken on vastly greater significance in the midst of the extremely tense standoff.

On October 26, just as the crisis was reaching its most dangerous point, the United States launched an unarmed Atlas ICBM from California. Both sides also proceeded with several previously scheduled nuclear tests during the crisis. Such activities might have been essentially routine under normal circumstances, but in the depths of a nuclear standoff, they were easy to mistakenly (or intentionally) misinterpret as cover for an attack. Under SAC's increased alert status, some safety and security procedures were disregarded, and nuclear-armed aircraft and missiles were deployed without the usual safeguards preventing unintentional detonation. Such slipups increased the possibility of a catastrophic accident—if not an accidental attack on Cuba or the USSR, then possibly the unwanted detonation of a nuclear weapon on US soil.

Defensive systems meant to warn each nation of impending attack also generated false radar reports and misidentified aircraft detections. Around the same time that Khrushchev was broadcasting his conciliatory message to the world on Sunday, October 28, a radar station in Moorestown, New Jersey, picked up an apparent missile launch from Cuba against the United States. The radar operators alerted SAC, but no missile detonation followed the warning. Investigation revealed that a test tape had been running on the system and led to the mistaken detection.

Probably the most bizarre such incident during the Cuban crisis came when security guards at an air force base near Duluth, Minnesota, already keyed up on high alert against possible saboteurs or a direct attack, detected an intruder trying to get through the base perimeter fence. It turned out to be a restless black bear foraging for supper.[10]

Another disquieting fact that emerged years after the crisis was that small tactical nuclear weapons intended for use on the beaches against an invading army were present in Cuba. If US troops had attempted to land on the island, local commanders had the authority to stop them by nuking the beachhead. It would have only been a limited tactical use of nuclear weapons, confined to a small area and not involving civilians, but it would have crossed the invisible line between conventional and nuclear war, and placed the United States under irresistible pressure to respond in kind. Once the nuclear threshold was crossed by both sides, escalation to a full-fledged strategic nuclear war would have been practically inevitable.

The Cuban missile crisis continues to be one of the most intensely studied as well as debated military and political episodes in human history, and for good reason: it nearly ended human history. And while it is universally acknowledged to be the occasion in which nuclear war—at least the full-blown, all-out, apocalyptic variety rather than a more limited exchange—was most likely, similar occasions have transpired since then, most far less visible or widely known.

Chances and Accidents

Perhaps the most disturbing aspect of most of these other incidents is that they occurred wholly outside of public awareness. It's probably safe to say that during October 1962, few citizens of at least the United States or its allies were unaware that the end of the world was hanging over their heads. Conversely, more than once both during and after the Cold War, the world briefly teetered on the edge of nuclear war while no one save for a handful of military personnel and officials knew anything about what was happening.

Two examples occurred at the North American Air Defense Command (NORAD) headquarters based in Colorado that monitors and coordinates defense for the entire North American continent. In November 1979, a training exercise tape was mistakenly fed into NORAD's computer systems, giving false warning of a Soviet ICBM attack. The mistake was quickly discovered and rectified, but not before preparations and warnings for war began to go into effect. In June 1980, a flawed computer chip at NORAD caused a similar frantic response. In these and a series of similar episodes throughout the Cold War, a tightly wound response system had simply been reacting as it had been designed—almost automatically and instantaneously with a minimum of human intervention, because in a real war, there would be only minutes between warning and the decision to respond.

Other unfortunate episodes with nuclear weapons have centered not on misinterpretations or misunderstandings of possible imminent attack but instead on simple accidents and mishaps involving nuclear weapons, called "Broken Arrows" by the Pentagon. In 1961, the Strategic Air Command almost detonated an H-bomb on the United States itself. A B-52 bomber flying over North Carolina broke up in midair, dropping two hydrogen bombs north of Goldsboro. Declassified documents released in 2013 revealed that one of the bombs had come close to detonating.

In 1966, a B-52 on a routine SAC airborne alert mission collided with a refueling aircraft over Palomares, Spain. The B-52 exploded, and the four hydrogen bombs it was carrying fell to the earth, with one lost at sea until it was eventually recovered months later. Although the conventional explosives inside the weapons went off on impact and scattered radioactive material, no nuclear detonation occurred.

Two years later, the crew of another B-52 on airborne alert over Greenland was forced to bail out when a fire broke out on its plane. The bomber, again carrying four hydrogen bombs, crashed several miles from the Ballistic Missile Early Warning System radar station in Thule. Again there was no nuclear explosion, but the plutonium from the weapons contaminated a wide (though in this case fortunately unpopulated) area.

These incidents were so alarming in their possible implications and vivid demonstration of the dangers of keeping nuclear-armed aircraft constantly in flight that they soon convinced the United States to cancel the SAC twenty-four-hour airborne alert program in 1968. Yet that did little to diminish the threat of accidental nuclear war. The United States and USSR still had plenty of missiles sitting in silos on hair-trigger alert, ready to be launched within minutes of the appropriate order.

Sometimes the precarious dance along the precipice of nuclear catastrophe has involved a broader timescale than the endless yet limited minutes required to detect an attack and prepare to retaliate, or even the thirteen days of the Cuban missile crisis. During President Ronald Reagan's first term in office beginning in 1981, relations between the United States and USSR were perhaps at their highest ever levels of tension since October 1962,

Figure 8 Despite the dedication of nuclear forces such as the Strategic Air Command to peace and deterrence, their hair-trigger posture led to many dangerous incidents. (National Archives)

fueled by increasing Soviet belligerence (including its recent invasion of Afghanistan and the shooting down of a civilian Korean airliner) along with the provocative "evil empire" rhetoric of Reagan and his administration. Both sides were spending vast sums to increase and modernize their nuclear arsenals too, with the Soviets having eliminated the inadequacy that hamstrung them in 1962, and the United States determined to regain and maintain supremacy.

By 1983, elements of Soviet intelligence had convinced themselves that the United States was preparing to strike first, and every Western move was interpreted as confirmation of that nonexistent intention. When the paranoid Soviet government feared that a routine NATO military exercise code-named Able Archer, which included a simulated nuclear exchange in Western Europe, might be a cover for an actual US attack, Russian forces were placed on high alert, preparing to strike the first blow. Fortunately, calmer minds prevailed within Soviet inner circles that realized that Reagan's bellicose posturing against the USSR wasn't backed up by any actual US military action to indicate imminent war.

In all these various scenarios, beyond the technical aspects of the weapons involved and the military strategies in play, the common denominator is perception: how the intentions and motives of one side are perceived by the other, and the mistakes that occur when intentions and motives are misperceived as well as misunderstood. Too often, actions are interpreted, at least initially, through the lens of one's worst fears, using a worst-case scenario perspective. For example, Kennedy and his advisers had to assume the worst and think not about what Khrushchev *might* do but instead what he *could* do with missiles only ninety miles away from US shores. Similarly, when nuclear-armed enemy bombers stray over one's border, when one's radar network insists that a blizzard of ICBMs is coming, or when the other side is playing war games that involve moving military forces in threatening ways, it's only prudent to be prepared for the worst—because if the worst *is* true, there will be no second chance. Where nuclear weapons are concerned, the concept of "benefit of the doubt" does not exist.

As I will examine later, the issue of perception is also central to deterrence: the seemingly contradictory idea that the only way to prevent nuclear war is to prepare for it. Perhaps even more illogically, this particular notion not only instigated many, if not most, of the crises I've discussed in this chapter but stopped them from resulting in ultimate disaster too. Such conundrums again highlight the essential uniqueness of nuclear weapons—the fears, respect, and even bizarre enthusiasm that they inspire—because they *are* fundamentally different. I'll now take a closer look at the physical and scientific characteristics at the core of those differences, and what actually happens when a nuclear weapon explodes.

BRIGHTER THAN THE SUN:
WHAT NUCLEAR WEAPONS DO

Throughout the first few decades that followed their introduction in 1945, nuclear weapons held a greater presence in human consciousness than they do now in the twenty-first century. A significant number of people living in the world during those years had actually seen nuclear explosions with their own eyes, giving them a more authentic and visceral awareness of their power that is missing in those of us who have only experienced them through grainy newsreel footage and photographs.

When Kennedy and Khrushchev faced off in the Cuban missile crisis in October 1962, they did so as leaders who had witnessed Hiroshima, Nagasaki, and the hundreds of open-air tests strewn across the late 1940s into the early 1960s, as had the political leaders, strategists, and soldiers who originally conceived as well as promoted the concept of nuclear deterrence. In the final analysis, it wasn't some cool diplomatic calculation or game theory abstraction that convinced Kennedy and Khrushchev to pull back from the abyss. It was a primal, human understanding of the stark reality of what they were each threatening to do—an understanding that now seems conspicuously absent from the sensibilities of world leaders, who talk of nukes as mere abstractions, geopolitical bargaining chips, and diplomatic hole cards.

The survivors of Hiroshima and Nagasaki, the *hibakusha*, are steadily vanishing. So are the thousands of military veterans, technicians, and scientists who witnessed at least one of the more than 520 atomic detonations that have occurred in the atmosphere (or in a few cases, underwater and in outer space) between 1945 and 1980, when China finally moved its testing underground. As with the veterans of long-ago wars, their memories die with them.

For now, the world's nuclear powers continue to observe the limitations of the 1963 Limited Test Ban Treaty prohibiting tests in the atmosphere, space, or underwater as well as a mutually agreed-on moratorium on underground nuclear testing. The United States has not exploded any nuclear weapons since 1992, and the other acknowledged powers have done likewise. Newer entrants into the nuclear club such as Pakistan and North Korea haven't observed such restrictions, although they have at least kept their testing underground. But unless and until the 1996 Comprehensive Test Ban Treaty ever goes into actual binding effect instead of continuing as little more than a symbolic gesture of noble intentions, nothing prevents any of the nuclear states including the United States and Russia from resuming testing at any time, save for the considerable technical preparations and great expense it would entail.

Even if that happened, however, such tests would still be deep underground, out of sight and out of mind for most people. These unfortunate truths make it necessary to understand just what happens when a nuclear weapon fulfills its intended purpose and explodes. Fortunately, the more than two thousand test explosions under almost all conditions and environments, not to mention the Hiroshima and Nagasaki attacks, have provided a wealth of data that allow us to describe, if not actually witness, the singular phenomenon of a nuclear detonation—and thus speculate with a high degree of accuracy what would happen if it occurred in a modern city. I will next consider both scenarios.

The Abstractions

In chapter 1, I explained the basic scientific facts of nuclear fission and fusion, but stopped short of the final result of those processes when they are allowed to progress rapidly and uncontrolled: a nuclear explosion. The specific phenomena accompanying a nuclear detonation can vary in their details, depending on the type of weapon (fission or fusion), and whether it's detonated in midair, at ground level, underground, at sea, or in outer space. The design of the weapon and how it's delivered on a target can also be important factors, as is the immediate environment, including the various meteorologic conditions such as ambient temperature, humidity, pressure, and wind. But there are several major things that take place in every nuclear detonation.

When the radioactive heart or "pit" of a nuclear weapon goes supercritical, caused by a device called the initiator that rapidly injects a flow of neutrons to trigger the process, the nuclear fuel heats within milliseconds to a temperature of tens of millions of degrees, hotter than the sun's core. Everything in the immediate vicinity, including the components of the weapon, is transformed into a state of matter called a plasma, which instantly begins to expand in a spherical mass called a fireball. A broad spectrum of electromagnetic radiation streaks outward in all directions, consisting of unimaginably brilliant visible light, gamma rays, and X-rays. To human observers, this light is the initial manifestation of a nuclear detonation, and will instantaneously blind any person unfortunate to be looking directly at it when it appears. This flash travels at the speed of light, or approximately 186,272 miles per second.

Accompanying the light of the nuclear fireball is a thermal pulse—that is, heat—that completely vaporizes everything within it and everything it touches as it expands. Although it rapidly dissipates as it spreads outward, the heat is so intense that it will burn everything in the vicinity and up to miles away. Many of the casualties and much of the damage at Hiroshima and Nagasaki came not from the blast effects but simply from burns and fire.

The fireball cools and rises as it continues to expand because its initial density is decreasing, but the thermal pulse is immediately followed by the shock wave consisting of superheated and supercompressed air hurtling outward. The shock wave acts like the similar effect of a conventional explosion, yet is far faster and more powerful, razing everything within its reach, moving at supersonic speed as a wall of wind more destructive than a hurricane.

In addition to the heat and shock wave, the intense gamma and X-ray radiation generated within the explosion is spreading outward at light speed. Oddly enough, despite this intensity, few of the people immediately killed by a nuclear explosion are the victims of radiation simply because of the gruesome fact that they are already killed by the shock wave or thermal pulse before radiation can claim them. Moreover, like the initial light flash, this radiation flashes and is then gone, which is why it's termed "prompt" radiation. Rather, most lethal effects come later in the "residual" radiation as debris swept up in the radioactive cloud that follows the fireball becomes contaminated and falls back to the ground as fallout. Because a device detonated at or near ground level will suck up more dirt and debris, these detonations produce more fallout than bombs exploded in midair.

Except for the prompt radiation, the major phenomena of shock/blast and heat/thermal radiation are present in conventional explosions, although of course they are far, far more powerful in a nuclear explosion. And while both conventional and nuclear explosions do most, if not all, of their immediate damage from the shock/blast effect, the destruction wrought by the heat or thermal radiation of a nuclear weapon is proportionally greater than in a conventional weapon. Particularly in a city, the intense, widespread fires ignited simultaneously over such a vast area by a nuclear detonation are a major contributor to the number of casualties.

While it may be true that blast, heat, and even radiation are phenomena that can be encountered at any time, a nuclear explosion manifests them all in an extremely brief time period and at a scale more extreme than in any other possible event. We may experience the devastating force of strong winds and kinetic energies in the natural world in the form of hurricanes or tornadoes; large fires are routine; and exposure to a certain amount of ionizing radiation is an inescapable part of daily life. But nowhere else can we be subjected to all of them simultaneously and at their greatest magnitude except in a nuclear explosion.

In chapter 1, I discussed the fundamental differences between a fission explosion in an atomic bomb and a fusion explosion from a thermonuclear or hydrogen weapon, and while there are some differences in the overall effects that they generate, the basics are the same: blast, heat, and radiation. Of course, these effects will be orders of magnitude greater with a hydrogen weapon compared to a mere fission bomb, but at least for practical purposes, both fission and fusion explosions create the same types, if not the same degree, of havoc.

Other effects will depend on where the detonation occurs. A blast at ground level will create a huge crater, the size of which depends on the yield of the weapon. All the irradiated soil thereby excavated by the blast is hurled skyward, where it becomes part of the radioactive cloud and eventually comes back to the earth as fallout. An underwater burst vaporizes an immense volume of water in a giant bubble, with the shock wave creating a swiftly moving base surge radiating outward that can engulf and sink ships, inundate islands, or devastate a coastline. If the water isn't too deep, a crater can be carved out in the ocean floor; many remain from the H-bomb tests in the Pacific during the 1950s.

One signature of a nuclear weapon known to everyone is the mushroom cloud, now invariably associated with the atom. Such clouds, however, aren't unique to nuclear explosions and can be seen with conventional explosives under the proper conditions. The mushroom forms when the extremely hot initial fireball of the explosion rises into the air, drawing the much cooler air below up into the expanding cloud. Along with this rush of incoming cooler air comes dirt and debris, which is sucked up into the expanding cloud and circulates within. This forms the top of the mushroom, with the debris still being drawn upward to form the stem. (In some of the many films of test explosions, one can actually see the stem forming and rising up to join with the top of the cloud to form the characteristic mushroom shape.)

The speed with which all this happens, and the particular size and shape of the cloud, will vary depending on the bomb and local atmospheric conditions, but some type of mushroom cloud almost always appears with a nuclear weapon detonated at or relatively near the ground. With extremely powerful thermonuclear weapons, other atmospheric phenomena are generated such as ice cloud condensation caps forming high above the bomb fireball and cloud. In one of the most telling illustrations of the power of hydrogen weapons, some test films actually show natural cloud formations being abruptly pushed aside, as if the thermonuclear cloud is punching a hole in the atmosphere.

These striking phenomena are missing when nuclear explosions occur at extremely high altitudes or outside

Figure 9 A 1955 nuclear test in Nevada. Note the debris stem rising from the ground to meet the fireball, forming the characteristic mushroom cloud. (National Nuclear Security Administration)

the atmosphere in outer space. From 1958 through 1962, the United States and USSR conducted about twenty such tests—some lofted with balloons, and others with warheads hurled into the heights on missiles—thus offering each nation the convenient benefit of testing its nuclear weapons and ballistic missiles at the same time. These bursts, occurring in an environment of little or no atmosphere, showed considerable differences from those nearer to the earth, with far less, if any, blast or shock wave effect and greater radiation effects, especially in the X-ray and visible light spectra.

At the proper altitudes, nuclear explosions can also interact with the earth's natural magnetic fields and radiation belts, creating striking auroral effects. The effects of the largest nuclear explosion in space, the US 1.4-megaton Starfish Prime test in July 1962, were seen and felt practically all across the Pacific Ocean. Another notable effect of high-altitude and space bursts is a strong pulse of electromagnetic energy known as the electromagnetic pulse, capable of temporarily or, in some cases, permanently knocking out radio and radar communications, and sometimes local power grids, all of which were seen with the Starfish Prime shot.

What happens when all these effects take place not in the benign and neutral setting of a testing ground but rather in the heart of a major city teeming with human beings?

The Practicalities

A nuclear blast packs within it every variety of natural disaster at its worst, within the smallest span of time. Though imperfect and subjective, one way to get a sense of

A nuclear blast packs within it every variety of natural disaster at its worst, within the smallest span of time.

the phenomenon is through the experience of those who have personally witnessed it.

We have many such accounts, from those who experienced the first atomic bomb at Trinity, the survivors of the atomic attacks on Japan, and those who have observed the nuclear tests of the Cold War. William L. Laurence, the *New York Times* science reporter who was the only journalist permitted at the Trinity test, authored a seminal description:

> And just at that instant, there rose as if from the bowels of the earth a light not of this world, the light of many suns in one. It was a sunrise such as the world had never seen, a great green super-sun climbing in a fraction of a second to a height of more than 8000 feet, rising ever higher until it touched the clouds, lighting up earth and sky all around with a dazzling luminosity.
>
> Up it went, a great ball of fire about a mile in diameter, changing colors as it kept shooting upward, from deep purple to orange, expanding, growing bigger, rising as it was expanding, an elemental force freed from its bonds after being chained for billions of years. . . .
>
> It was as though the earth had opened and the skies had split. One felt as though he had been privileged to witness the Birth of the World. . . . On

that moment hung eternity. Time stood still. Space contracted to a pinpoint.[1]

Still, Laurence was observing the bomb from a safe and comfortable distance, not up close and personal, like the people of Hiroshima and Nagasaki. The first-person accounts collected by John Hersey in his landmark work *Hiroshima* have lost none of their power in the decades since their 1946 publication in *The New Yorker*:

> A tremendous flash of light cut across the sky. Mr. Tanimoto has a distinct recollection that it travelled from east to west, from the city toward the hills. It seemed a sheet of sun. . . . [H]e felt a sudden pressure, and then splinters and pieces of board and fragments of tile fell on him. He heard no roar. (Almost no one in Hiroshima recalls hearing any noise of the bomb.)
>
> As Mrs. Nakamura stood watching her neighbor, everything flashed whiter than any white she had ever seen. . . . [A] shower of tiles pommelled her; everything became dark, for she was buried.[2]

That sudden flash of brilliant light is a consistent, almost universal part of all eyewitness accounts of nuclear explosions, as are a sensation of sudden heat on the skin, often described as "opening the door of an oven." Then comes the experience of the shock wave, at first almost felt subliminally in the body like the first rumblings of an earthquake, before swiftly rising into an unbearably loud, deafening noise, accompanied by punishing winds.

Because it would be intended to cause maximum destruction and loss of life, an attack on a major city would likely consist of an airburst, which spreads the nuke's destructive power over the widest possible area. The size of the initial fireball would depend on the weapon yield, ranging from tens of meters to several kilometers in radius, but everything within it would simply cease to exist, completely vaporized without a trace. Extending outward, the destruction would be almost total within several miles of ground zero (a term invented for nuclear testing, not originating with the 9/11 attack, as commonly believed). As previously noted, the immediate casualties would be victims of the blast and heat, with uncounted hundreds or even thousands trapped in the ruins of buildings.

Radiation casualties would follow, most of them from the initial radiation from the detonation. With an airburst, defined as a detonation in which the fireball doesn't touch the ground, the amount of fallout that people downwind of the attack would have to face would be limited, but residual radiation would be a major concern for rescue personnel within the attack zone.

For the first hours and days following the attack, the survivors would be largely on their own. Streets, roads, bridges, tunnels, and all access to the stricken areas would be destroyed or impassable because of debris or fire, making it impossible to reach survivors. Communications networks, including cell towers, fiber-optic lines, and the intricate infrastructure that ties together the internet, would also be damaged or destroyed within the attacked area, so cell phones and computers would be largely useless. Hospitals and other local medical facilities would mostly be wiped out, as would a great number of doctors, nurses, and other medical personnel.

If an attack were only limited to a single city or perhaps only a handful, aid would of course pour in from outside, although it would be greatly hampered by all the destruction as well as the remaining hazards of fire, collapsing structures, and radioactively contaminated areas. But in the face of a major attack involving the entire nation, there may be no "outside" to provide immediate, significant assistance. Even without such exigencies, the lack of help for survivors will contribute to more casualties beyond the initial victims. Consider the examples of New Orleans after Hurricane Katrina or Puerto Rico after Hurricane Maria. And in these and other similar natural disasters, there was plenty of warning and time to prepare (even though the warnings were ignored and preparations were not made). A nuclear attack on a city, whether a military or terrorist one, would likely come with little or no warning, and encompass far more destructive power in far shorter a time than a hurricane or any other natural catastrophe.

The distinction between a terrorist nuclear attack and the military variety is worth examining. I will look at nuclear terrorism in more detail in chapter 6, but in the context of considering the damage resulting from both types of attack, the terrorist possibility would likely be less severe and cause fewer casualties. Modern military nuclear weapons and the tactics for their use are highly sophisticated, whereas a terrorist device would most likely be relatively primitive, and thus far less powerful and efficient. Its crude design would be somewhat offset by the fact that a terrorist bomb would probably be detonated on or near the ground, exchanging greater explosive yield for more radioactive fallout. Either way, such differences won't matter much to anyone within several miles of ground zero.

One approach that a dedicated terrorist group might take to circumvent the difficulties of building an actual working nuclear weapon is a "dirty bomb": a conventional bomb that's laced with radioactive material. It's a reasonable alternative because it avoids any worries about assembling a supercritical mass in exactly the right way with precise control. The only problem is getting a fair amount of radioactive material of any type—something that's fairly common and easy to obtain, unlike weapons-grade uranium or plutonium. Low-grade radioactive material is routinely used for medical and industrial applications. Scattered over a populated area by a typical bomb of the type often built by terrorists around the globe, the radioactive material would be readily detectable by first responders, and even if the radioactivity were too low to cause any actual effects on people, the resulting psychological impact and panic would cause disruption far out of proportion to the damage actually caused by the bomb. Considering such advantages and the low cost of such a weapon, it's somewhat surprising—if also relieving—that we've yet to see one employed.

It's impossible to get too specific about the results of any particular nuclear weapon on any particular target; the variables of atmospheric conditions, geography, time of day, available resources, and a thousand other factors make it too complex a question. Nuclear historian Alex Wellerstein, however, has provided one way to get some idea of what might happen if your hometown were nuked. His NUKEMAP website (https://nuclearsecrecy.com/nuke map/) allows you to simulate the effects of various weapons on any location. Ultimately, though, the details don't matter. Whatever the specifics, any nuclear detonation in any city, town, or area populated with human beings would be a catastrophe of unimaginable proportions—an event with repercussions that would reverberate far beyond the immediate damage and deaths into an uncertain future.

In a real sense, at least when considering the hypothetical scenario of a single nuclear attack on a single city, the psychological and social impact would be far more significant than the hundreds of thousands of deaths and billions of dollars of damage that would result. We had something of a preview of this in the aftermath of the 9/11 attacks. Not only did that national trauma lead to US involvement in new wars abroad, but it brought stronger, more intrusive security measures, surveillance programs, and curtailment of traditional civil liberties at home, all largely condoned and even welcomed by US citizens terrified of any further terrorist attacks on US soil. All this came as a result of the approximately three thousand people who died on September 11, 2001—a mere fraction of those who would perish in any nuclear attack. It's not hard to imagine that in the aftermath of such a disaster, panicked by the specter of perhaps more to come, people in the United States would readily submit to draconian measures to guarantee their safety, up to and including martial law. As with 9/11, other social and economic impacts would be felt worldwide and reverberate for years afterward.

As I've mentioned earlier in this book, the sociopsychological factors and perceptions surrounding nuclear weapons are a major reason to consider them separately from other weapons. Chemical and biological weapons are also regarded as fundamentally and qualitatively different from conventional types—set apart into a class of their own with nuclear weapons as "weapons of mass destruction." Yet nuclear weapons don't seem to evoke the same visceral sense of horror and revulsion, even though they're far more efficient in killing people. They remain much more of an abstraction, at least for anyone except the *hibakusha*. And of course, those who were vaporized into nothingness in Hiroshima and Nagasaki can tell us nothing since they no longer exist.

It's this abstractedness surrounding nuclear weapons that ever since their invention, have made them a subject for endless theorizing, conversations, and intellectual debates. Unlike any other subject with such all-encompassing significance to humanity, much of this discussion, planning, and policy making has taken place well out of the awareness of the public, and has therefore affected much of the political, military, economic, and social trends and events in human societies from well behind the curtain, like the plotting of a malign Wizard of Oz. I've examined the essential facts of nuclear weapons; I'll now look at some of the political and military strategies, rationales, plans, and policies for dealing with, controlling, and using nukes that have arisen since even before Hiroshima.

DETERRENCE AND DOOMSDAY: NUCLEAR THEORIES AND TREATIES

Although at first it was limited to a quite small and exclusive group, discussion and agonizing over the use of nuclear weapons started long before anyone knew that an atomic weapon could actually be built. But after Hiroshima and Nagasaki, when the Bomb had proven itself as a viable weapon, political leaders, military men, and even some scientists began seriously thinking about how to use nuclear weapons on US enemies, and how to prevent them from being used on the United States. The shape and details of such discussions evolved over the years under the influences of political priorities, technological advances, and strategic objectives, but always boiled down to these two essential concerns: a basic conflict of wanting to have one's nuclear cake and eat it too.

Supremacy and Competition

At first, for the United States at least, matters were simple. Since only the United States had the Bomb, it held unquestioned dominance and could dictate terms to the rest of the world, with only its traditional sense of morality and democratic ideals to restrain it. Some officials even advocated making full use of the advantage by launching a preventive or preemptive war against the Soviet Union, removing it decisively as a threat for the foreseeable future. Fortunately, such proposals were never taken seriously by the leaders who would have had to make the decision.

By 1950, with the Soviet bomb a reality and West fighting East in Korea, the basic doctrinal outlines of the Cold War had formed, centered around a concept called deterrence. In its most elemental form, we learn about deterrence on the playground. If someone hits you, you hit them back; if you hit first, then you expect them to hit back. From this basic principle, it follows that the bigger and stronger you are, the less likely it is that someone will want to hit you for fear of the inevitable retaliation that they will face. As a corollary, the bigger and stronger you are (or are perceived to be), the more freedom you might feel to throw your weight around by bullying others, if you're so inclined.

On the playground of international relations, bigger and stronger translates to military power, and in nuclear terms, that means ever-greater numbers of ever-better nuclear weapons. In the theory of nuclear deterrence, your enemy is deterred, or discouraged, from launching a nuclear attack on you because they know that it will cause overwhelming retaliation on themselves. The downside is that this also provides a strong incentive for them to overcome that handicap by building enough weapons to overcome or preferably preempt your retaliation by knocking it out before it can be launched. This, in turn, forces you to build even more weapons to avoid losing your strategic edge, thus embarking on an ever-expanding arms race with both sides chasing the chimera of absolute dominance.

Why not settle for a rough parity on both sides rather than constantly striving for superiority, spending more and more money while straining one's economy? That would seem to be the logical solution, but it fails to account for human nature. First of all, the side with nuclear superiority—throughout the 1950s, the United States—would have to simply allow its opponent to "catch up" with it, trusting them to stop when some mutually defined level of parity was achieved. Second, both sides would have to trust each other enough not to cheat by trying to secretly obtain some edge by building and hiding more weapons, or developing some other military advantage that its enemy doesn't have. As we will see, the treaties and diplomatic efforts to slow the arms race that began later in the Cold War were essentially attempts to address these problems.

Deterrence was first officially articulated in the Eisenhower administration's policy of "massive retaliation" against any Soviet nuclear attack. This came as part of Eisenhower's "New Look" defense posture, which chose to rely on nuclear versus conventional weapons as the United States' main bulwark against the Soviet threat. The New Look also had an economic rationale for the budget-conscious Eisenhower. While nuclear weapons were certainly expensive, it was still relatively cheaper to maintain an adequate nuclear deterrent than vast conventional forces deployed all around the globe.

The New Look and doctrine of massive retaliation built up the Strategic Air Command at the expense (and considerable consternation) of the army and navy, both of which devoted great effort throughout the 1950s to find some way of grabbing a bigger slice of the nuclear pie. At the time, however, the only means to deliver nuclear weapons was by aircraft, and later, long-range missiles—tools that fell solidly under the aegis of the air force.

The Soviet Union, of course, wasn't sitting still as the United States continued to build and test ever-larger, and more devastating hydrogen bombs, and as General LeMay's bombers continued to conduct mock atomic war on US cities and his reconnaissance planes probed the edges of the Soviet Empire. It steadily built up its own arsenal, announcing its existence through test shots and blustery propaganda. Because its weapon designs tended

Figure 10   Long-range bombers such as the Boeing B-52, fully armed with nuclear weapons, continually patrolled the skies during the height of the Cold War, always poised to retaliate if the United States were attacked. (US government)

to lag behind more sophisticated US weapons to a certain extent, a technological reality that made Russian bombs somewhat heavier and larger, the USSR was forced to find ways to circumvent the problem, which meant putting bombs atop missiles rather than carrying them in the belly of airplanes.

By the time the Soviets demonstrated the prowess of their missiles by using them to launch the world's first satellites, the game had suddenly changed substantially, and with it the strategic and military picture. A secret report commissioned by Eisenhower soon after Sputnik claimed that the USSR would have a missile force able to devastate the United States within only a few years, whereas the United States could barely get a rocket off the ground. Russian ICBMs would be able to totally overwhelm LeMay's now-obsolete force of almost two thousand bombers. Because of this "missile gap," the next few years would be "years of maximum danger," in which the risk of a Soviet first strike against the United States was greater than ever before. Deterrence was starting to look decidedly shaky, at least from the point of view of the United States.

But deterrence is also a matter of perception and psychology, not merely technology and the size of one's arsenal. As events proved, in particular the U-2 spy plane flights that were providing accurate information on Soviet capabilities, the supposed missile gap favored not the USSR but instead the United States. Spurred by its own anxieties, the United States quickly surpassed Soviet missile technology and achieved a major breakthrough with the development of missiles that could be launched from submarines, ensuring the survival of a deterrent capability even in the worst-case scenario of an attack that wiped out the entire land-based US bomber and missile force. This led to the concept of the strategic "triad," a three-pronged retaliatory capability of ICBMs, submarine-based missiles, and the trusty manned bombers. One or even two parts of the triad might be taken out, but to eliminate all three would be practically impossible—thus preserving deterrence.

For some strategists, the emergence of these new weapon technologies also meant that nuclear war no longer had to be an all-or-nothing proposition, a choice between massive retaliation or utter destruction. Perhaps now it would be possible to have "flexible options," even to the extent of actually placing limits on nuclear war. For President Kennedy and his secretary of defense, Robert McNamara, that seemed attractive. Kennedy didn't relish the idea that if, for example, a Russian missile were launched by accident and destroyed one US city, he would have no other option but to launch a massive attack on the USSR.

The concept of flexible response led to the idea of differentiating between possible targets by categorizing them as "counterforce" or "countervalue." Counterforce targets were strictly military: missile bases, air bases, enemy troop concentrations, ships and submarines, and aircraft approaching one's borders. Countervalue targets were civilian: cities and industrial centers.

The point was to attempt to restore, however imperfectly, the old idea that warfare should involve only combatants—that is, soldiers as opposed to innocent civilians. In practice, it had always been an artificial and unrealistic distinction since the beginning of organized warfare, but one that had at least been tacitly acknowledged until the world wars of the twentieth century. Perhaps now it could provide a way to limit the devastation of nuclear warfare. Under this philosophy, if nuclear war broke out, counterforce targets would be hit first, thereby limiting civilian casualties. Countervalue targets could be saved as bargaining chips—hostages, in effect—to provide a means for one side to force the other into acquiescence. Only if the war continued would cities be hit. It all sounded quite rational.

The problem, of course, was that human beings, particularly when acting in large groups such as nations and armies, are often irrational. There was no way to guarantee that the Soviet Union would behave according to McNamara's reasoning, nor that the United States would stick to such rationales once the missiles started flying and millions of its own people were being exterminated.

A Shaky State of Affairs

By the mid-1960s into the 1970s, the USSR had achieved a rough strategic nuclear parity with the United States, fully adopting the humiliating lessons of the Cuban missile crisis. Both sides had large numbers of highly accurate ICBMs based on land and submarines; both sides had elaborate early warning and communications systems to control

The problem, of course, was that human beings, particularly when acting in large groups such as nations and armies, are often irrational.

their forces; and both sides maintained fleets of bomber aircraft to serve as a second-strike resource. Each side also maintained spy satellites and other reconnaissance technology, including the ability to collect and analyze each other's various electronic signals, so they were able to keep tabs on each other, and have at least an approximate notion of their enemy's strengths and weaknesses.

A sort of stability had been achieved via MAD (mutual assured destruction). Earlier concepts of deterrence held that the mere threat of having as little as one or two nuclear weapons exploding on one's country was enough to forgo any thoughts of striking first. MAD took that thought to its logical conclusion. Now any attack by one nuclear power on another would mean not just retaliation but also the total destruction of the attacker's country and society for all practical purposes. What had been impossible or at best only an unlikely possibility for years was a definite prospect now that both East and West possessed such broad nuclear capabilities.

Yet rather than settle for this uneasy status quo, both the United States and USSR continued to vie for a decisive edge (while still extolling a desire to maintain stability by "preserving" MAD). One way to do that was by building more warheads, bombs, missiles, planes, and submarines, which each side proceeded to do as best as it could, despite the great economic strains and sacrifices it entailed. Until the mid-1970s, the United States was hampered by the interminable Vietnam War, which diverted not only money but also military resources from the nuclear mission. When tactical fighter-bomber aircraft proved inadequate to the rapidly expanding task of bombing the enemy, the B-52 bomber, designed and operated until then strictly as a strategic (read: nuclear) aircraft, was pressed into service to, as LeMay famously remarked, "bomb North Vietnam into the Stone Age" with conventional high-explosive bombs.[1]

Meanwhile, the Soviet Union spent vast sums to maintain and expand its own nuclear arsenal, without regard to the ways in which such spending was damaging its civilian economy and denying its own citizens basic necessities. But as it had been since the onset of the Cold War, technological innovation offered another path to achieving nuclear supremacy. Two major developments arising in the 1960s promised this goal—one defensive in nature, and the other offensive. Both would cause great anxiety for strategists, at least when the other side began working on them. Each development challenged on a fundamental level the supposedly sound foundations on which deterrence and MAD rested.

Measures and Countermeasures

One of the basic (as well as contradictory) tenets of deterrence is the idea of vulnerability. You understand and accept that you are subject to retaliation by your enemy if you choose to attack first, and that knowledge of your own vulnerability keeps you from doing anything so foolish. In effect, you believe that anything you might gain by attacking your adversary isn't worth the inevitable price that you would pay, and therefore you refrain from attacking.

But what if that changed? What if you came to believe that you *could* strike first, or take some other action that would invite attack, without paying that price? Another factor in deterrence is the general belief that nuclear weapons are so powerful that there's no possible defense against them. This idea is buttressed by the fact (or belief) that even one striking you is one too many—a distinction that conventional weapons don't have. Yet what if you could defend yourself against nuclear attack?

Offense and defense are fundamental concepts of military strategy, but nuclear warfare was generally considered the one exception. Because nuclear weapons are so terrible, the only defense against a nuclear attack is preventing it from happening in the first place, which means being strong enough to keep anyone from attacking you. If a military defense against nukes actually were possible, however, then militarily speaking, nuclear warfare suddenly moves from being inconceivable to being little different from any other type of warfare. And if that happens, the power of deterrence is fatally weakened, if not abolished.

The idea of nuclear war defense had existed from the early years of the Cold War, but because at that time nuclear attack meant enemy bombers over your country, it was based on air defense: shooting down the bombers before they could reach their targets. The experience of World War II had amply demonstrated that there was no such thing as a perfect defense against airpower. It was impossible to shoot down every attacking plane; some would always get through. If those planes carried nuclear weapons, that was enough to invalidate the concept of defense against nuclear attack and hence preserve deterrence. Against nuclear weapons, only a perfect, and thus unattainable, defense is enough. An imperfect defense can still save many lives and so has value, but can't prevent catastrophe.

In the 1960s, however, with the advent of missiles and more sophisticated radar tracking technology came the concept of antiballistic missiles (ABM), or shooting ICBMs down before they could hit you. Both the United States and USSR started developing such systems, despite the fact that on a strictly technical level, intercepting and shooting down a missile traveling at thousands of miles an hour is a far more complicated task than shooting down an aircraft. That didn't stop both nations from deploying rudimentary ABM systems without regard to their many imperfections or the certainty that even if they worked flawlessly, they could not stop a full-scale attack.

Against nuclear
weapons, only a perfect,
and thus unattainable,
defense is enough.

As far as deterrence and MAD were concerned, the technical realities of ABM systems were less important than their perceptual effects. A nation with a working ABM system might feel confident enough that it could strike first, even knowing that the ABM system was imperfect and couldn't ward off massive damage from retaliation. As long as enough of its own nuclear forces remained, went the reasoning, the attacker could still prevail by threatening another attack on its now-decimated foe.

Thus the ABM issue spawned the second major destabilizing technological innovation: the multiple independently targetable reentry vehicle (MIRV). With MIRV, an ICBM bears not one but rather several nuclear warheads, each of which can be individually guided to a separate target. This increases the destructive power of each ICBM while immensely complicating any defense, which would now have to take out each individual warhead (a small, hard to detect object) instead of simply a single missile (far larger) or single warhead. To make matters worse, a MIRV can deploy dummy warheads or other decoys to further confuse detection systems, and create so many targets that an ABM system would be overwhelmed.

These two technologies, ABM and MIRV, created other problems. Strategists on both sides argued that they provided greater incentives to strike first as opposed to using nuclear weapons only in retaliation. If one side believes that it can hit first with relative impunity, that can lead it to actually do so since the side that gets in the first licks in a nuclear war will likely suffer less damage than the side that holds back.

Again, these issues highlight the essential role of psychology in dealing with nuclear weapons and warfare. Humans often tend to base their beliefs on fear, worst-case scenarios, and hypothetical situations rather than the available facts, and rush to judgment and action instead of waiting for better information. Yet the reluctant acknowledgment of a rough parity between East and West, and ultimate impossibility of achieving any lasting supremacy, led both sides to seek another way out of the shaky state of MAD-based deterrence for the indefinite future.

Sitting down at the Table

Ever since the initial efforts at international agreement and control had failed at the dawn of the nuclear age, the United States and USSR as well as the allies that followed them into the exclusive club of nuclear powers on one side or the other had more or less settled into a comfortable us versus them mentality, believing that any kind of meaningful dialogue and agreement was out of the question. Some on both sides believed this so fervently that they insisted that World War III was inevitable, and thus there was no alternative but to somehow make certain that their side prevailed, whatever the cost.

But the irresolvable ABM and MIRV dilemmas, along with more practical concerns such as the sheer expense, forced the attempt to strike some kind of diplomatic understanding. The test moratorium of the late 1950s and resulting 1963 Limited Test Ban Treaty were the first steps, but these only addressed nuclear testing—a side issue at best. They did nothing to control or reduce nuclear weapons themselves. Neither did the 1968 Nuclear Nonproliferation Treaty signed by all five then-extant nuclear powers and a large number of other nations, with the latter pledging to forgo seeking nuclear weapons and limit themselves to peaceful nuclear technology, and the former promising to prevent the spread of weapons to other countries and work toward disarmament. While this treaty has been largely effective in limiting, if not completely preventing, the unchecked expansion of the nuclear club, the signatory members of that club have been lax in fulfilling their obligations under the treaty to eliminate their own arsenals.

The groundwork had nevertheless been set by these earlier initiatives for more serious work toward actually reducing and limiting nuclear weapons. In 1972, after extensive and difficult negotiations, the United States and USSR signed the Strategic Arms Limitation Talks (SALT) treaty and the ABM Treaty, the first serious arms control agreements. SALT placed some limitations on several types of missiles and other weapons systems, while the ABM Treaty limited both sides to one ABM site each, with the aim of preventing an unbridled ABM race. In reality, neither side was making any significant concessions that would seriously affect its strategic capabilities, but whether acknowledged or not, that was partly the point. It was enough to restore some sense of stability and preserve the deterrence status quo. The mere fact that both sides were talking to each other and addressing existential questions, and therefore acknowledging a certain degree of respect for each other, was at least a positive and hopeful development.

Many military leaders remained resistant to the entire idea of MAD. For them, accepting and, worse, allowing for the essential vulnerability of one's own nation seemed too much like admitting defeat, which was an inexcusable weakness. Surely there still existed some options for using nuclear weapons, not simply endlessly stockpiling and then guarding them while simultaneously intending never to release them. The motto of the Strategic Air Command might be "Peace Is Our Profession," and no doubt most of the men and women of SAC truly supported that ideal, but for some military strategists, it was foolish and wasteful to have such capability with no opportunity to exercise it.

The New Years of Maximum Danger

In the 1980s, after some years of military budget cutting and reductions in the aftermath of the Vietnam War, the Pentagon's fortunes changed radically under the Reagan administration. There was no more talk about détente or arms agreements. Instead, a massive expansion of defense spending ensued, with grand plans for new and ambitious weapons including ICBMs, strategic bombers, nuclear submarines, and various types of warheads and bombs for them all to carry. Added to this was the revival of, or perhaps sequel to, flexible response. Again fashionable was the idea that nuclear war was not an all-or-nothing proposition, a stark choice between the end of the world or abject surrender, but rather a viable option. MAD was an outmoded concept, argued these new Cold Warriors. Instead, it was possible not only to limit but actually win a nuclear war too. Military budgets and strategic planning began to reflect these new ideas.

Such notions might have excited strategists and generals, but they alarmed many others, most notably large segments of the public that would bear the brunt of nuclear war, limited or not, while those who started it snuggled safely in luxurious secret underground shelter complexes. Relations between the United States and USSR had already deteriorated sharply by the end of the 1970s with provocative moves on both sides, particularly in 1979 with the

Soviet invasion of Afghanistan, which killed the recently concluded SALT II agreement. The escalation of bellicose rhetoric and increased military spending by both East and West only made things worse.

A new wave of grassroots antinuclear campaigning sprung up in the United States and with a particular urgency in Western Europe, which was generally expected to be the place where World War III would begin with Soviet tanks pouring through the Fulda Gap into West Germany. Activists argued that if such a nightmare ever transpired, triggering a tactical nuclear response from NATO forces, then it would be Cologne and Hamburg and Brussels that would be wiped out first, rather than New York or Philadelphia or Washington, DC. For the people of Western Europe, nuclear war had a greater, more horrific immediacy than for the citizens of US cities.

As the United States and USSR continued their saber-rattling, the early 1980s saw some of the largest political demonstrations in history, in Washington, London, Berlin, and other Western cities. The new generation of organizers rallied around a new concept called nuclear freeze, which argued not for disarmament, reduction, or the elimination of nuclear weapons but simply a halt on building more—a "freeze" at the current levels. This, advocates asserted, would reduce the tensions of an ever-escalating arms race and allow a sense of stability to settle in from which serious negotiations could then proceed.

Deterrence wouldn't be threatened since no one would be giving up anything.

Providing more support and force to the peace campaigns was new research by scientists (among them astronomer Carl Sagan) from various disciplines, including climatology, meteorology, and planetary astronomy, that seemed to indicate that a large-scale nuclear war would cause damaging environmental effects so severe that a "nuclear winter" would ensue. The sudden injection of massive amounts of soot, smoke, and debris into the atmosphere from the fires resulting from hundreds or thousands of nuclear detonations would lead to a global drop in temperature that would alter the earth's climate for an indefinite period.

Based on new climate-modeling techniques that incorporated observed atmospheric effects from previous H-bomb tests, results of World War II aerial attacks on cities, and even atmospheric phenomena on other planets, the nuclear winter theory was alarming, yet also subject to a great deal of uncertainty. But for the public, it added a new dimension of horror to nuclear war and provided a strong riposte to those government officials who blithely insisted that practically anyone could survive a nuclear war as long as they had shovels to dig out a decent shelter for themselves (part and parcel with the return of flexible response doctrines, talk of civil defense measures and their efficacy found new life in this period). With nuclear winter, even if one were lucky enough to survive the war, they would be faced with trying to live in a cold, twilight purgatory with neither sunlight nor food.

All of which made the simple and powerful idea of the nuclear freeze even more appealing to many people. It seemed a rational and yet not too extreme first step to derail the inexorable spiral toward nuclear war. Predictably, the nuclear freeze movement met with strong opposition from official and military circles, where it was argued that a freeze would lock in strategic forces at dangerously unstable levels (which begged the question of what was preserving stability already, and if one side already had a decisive advantage, why they simply hadn't attacked yet), and that if one side cheated and continued to build weapons, it would soon prevail (a contention perennially deployed against almost every treaty since the start of the nuclear age).

Looking for the Exit

The nuclear freeze was beginning to gain traction beyond the activists, and serious interest and support within the US Congress and other political circles, when it was essentially defused by a Reagan administration initiative that shocked hawks and doves on all sides by apparently undercutting all their arguments. On March 13, 1983,

President Reagan delivered a speech to the nation in which he presented "a vision of the future which offers hope," proposing the development of a dedicated ballistic missile defense system that would "render these weapons impotent and obsolete." His Strategic Defense Initiative (SDI) would create a system that could "intercept and destroy strategic ballistic missiles before they reached our own soil or that of our allies."[2]

Even though it was almost immediately and somewhat derisively dubbed "Star Wars" by the press and public, the SDI was a brilliant political move. Reagan and his administration had been criticized as well as condemned by both the Soviet Union and US citizens since taking office as inveterate warmongers and militarists pushing the world toward nuclear war, escalating tensions with "evil empire" rhetoric while doing absolutely nothing to work toward disarmament. Now Reagan had pulled the rug out from such criticisms with one speech. He was arguing not for offense but rather for defense, not for building more weapons, but for building a means to protect against them.

Many within Reagan's own administration were shocked and dismayed by his proposal, or didn't take it seriously, but Reagan himself was quite sincere. Despite all the aggressive anti-Soviet rhetoric of his first years in the White House, he had become horrified on learning the realities of nuclear weapons and plans for their use, and became committed to finding some means to control, if not completely eliminate, them. Despairing of finding some ultimate modus vivendi with the Soviet Union, he saw another path when certain influential and politically active scientists, most notably Manhattan Project veteran and supposed "father of the H-bomb" Teller, convinced him that a technical solution was possible. With his trademark passionate, if not always practical, enthusiasm, Teller assured Reagan that technology had advanced to the point where it was definitely possible to develop a system capable of detecting, intercepting, and destroying enemy ICBMs in flight, and doing so flawlessly—or at least flawlessly enough that an attacker couldn't be assured of launching a successful and decisive first strike.

Many other scientists maintained just the opposite. Intercepting a single ICBM in flight was a problem far beyond any current or foreseeable technology, much less warding off a massive attack involving hundreds or thousands of missiles. It was a matter of not simply technology but instead basic physical principles. And even if it were eventually possible to "shoot down a bullet with a bullet," as the problem was characterized, it would always be possible to defeat any defense system merely by either deploying cheap decoys to fool it or overwhelming it with more targets than it could handle.

Teller and other SDI scientists dismissed such objections, and got to work with Reagan's blessing and virtually unlimited funding. But the technical challenges weren't the only obstacle to a workable SDI system. The development of a strategic missile defense was also in violation of the ABM Treaty and thus inherently destabilizing. SDI advocates might have been preaching peaceful defense in principle, but they were exacerbating, not reducing, international tensions for all practical purposes.

Moreover, some strategists pointed out that the SDI struck at the very heart of deterrence by undermining vulnerability—the same concerns that led to the ABM Treaty. As Soviet leaders noted at the time, it was all well and good to talk about strategic missile defense, but what would stop the side that had it from attacking its opponent? Instead of preserving peace, it encouraged aggression. Soviet anxieties about the SDI were real, and heightened because the USSR knew that it could not possibly hope to develop a comparable system on its own with its available technical and financial resources. A number of post–Cold War historians contend that the attempt to counter the SDI by building ever-more weapons and spending more money was a major factor leading to the collapse of the Soviet Union, whose economy simply couldn't sustain such effort. In fact, the refusal of the United States to forgo the SDI probably cost the world its only real chance to date of completely abolishing nuclear weapons.

That came in 1986 at Reykjavík, Iceland, when President Reagan and Soviet chair Mikhail Gorbachev met for a summit meeting. By this time, relations between the

United States and USSR had warmed to some degree, with both sides talking more about understanding rather than continuing to threaten each other. Gorbachev was instituting reforms within his own country offering more openness and economic freedom, and Reagan's attitude toward the USSR and Communism had softened and become more nuanced. So for a brief moment in Reykjavík, Reagan and Gorbachev almost agreed on something that for forty years previously had been unthinkable: complete and mutual nuclear disarmament over a period of ten years. Both leaders were quite sincere, with Gorbachev making the proposal and Reagan agreeing in good faith.

It seemed too good to be true, and it was. Gorbachev's one condition was that the United States, in the spirit of the ABM Treaty, had to limit its SDI program to the laboratory, without any testing or deployment. Reagan was unwilling to agree, insisting that the SDI posed no threat to the Soviet Union because the United States had no intentions of attacking. Reagan and Gorbachev were unable to reach a compromise, and the opportunity to abolish nuclear weapons slipped away, much to the regret of both men.

All wasn't lost, however. The rapport between Reagan and Gorbachev along with the resulting goodwill from the Reykjavík summit led to a major arms treaty in the following year, the Intermediate-Range Nuclear Forces Treaty, which eliminated all ground-launched nuclear missiles

Reagan and Gorbachev almost agreed on something that for forty years previously had been unthinkable: complete and mutual nuclear disarmament.

Figure 11 A glum President Reagan and Secretary Gorbachev head home from their 1986 Reykjavík summit after failing to reach an agreement to totally abolish their nuclear arsenals. (CTBTO Photostream)

with ranges from about three hundred to thirty-five hundred miles. This included most of the weapons that had been causing such consternation for NATO and Warsaw Pact nations on the European continent, and represented a substantial agreement in arms control. For the first time, an entire category of nuclear weapons had been abolished. It may not have been total disarmament, but it was an encouraging step in that direction.

Before the Cold War ended with the dissolution of the Soviet Union in December 1991, another major treaty was concluded with the Strategic Arms Reduction Treaty

(START I). Unlike the previous SALT treaties, which per their name only limited weapons, START required both the United States and USSR to actually reduce their arsenals to approximately six thousand nuclear warheads and sixteen hundred delivery vehicles (missiles and aircraft) each. A START II agreement was signed in 1996 by the United States and Russian Federation that incorporated even greater reductions. In 2010, the New START treaty was signed.

Old and New Threats

With all its inherent contradictions, absurdities, and conundrums, and despite being sorely tested on many occasions almost to the breaking point, deterrence had apparently survived intact for the first half century of the nuclear age. Now, though, the fundamental parameters within which it had been operating had changed completely. The breakup of the Soviet Union temporarily increased the number of nuclear powers in the world, as the former Soviet republics of Ukraine, Belarus, and Kazakhstan inherited the weapons based on their territories.

Two US congressmen from both sides of the aisle, Republican Richard Lugar and Democrat Sam Nunn, won approval and funding for a concerted joint effort with the Russian government called the Cooperative Threat

Reduction program to secure what were now known as "loose nukes." The arsenals of Belarus, Ukraine, and Kazakhstan were destroyed, or bought and removed by the Russians, and the nuclear materials were diverted to civilian use under US supervision.

That restored the standing membership of the nuclear club as well as the essentially bipolar balance between the United States on one side and Russia on the other as the two major players. But in the shifting grounds of the post–Cold War world, new concerns soon arose and new threats began to emerge. If the dangers of a global nuclear conflict between East and West had greatly diminished (if not entirely disappeared), the possibility of nuclear war on a smaller scale increased between nations that either already had nuclear weapons or were striving to acquire them. Formerly, such ambitions were restrained either tacitly or directly by the United States or USSR through political influence or military threat, but that source of stability was now gone. Most of the world's nations had previously aligned themselves either with the West or East, and more or less followed the dictates of the major powers, but now countries such as India or Pakistan felt more freedom to assert their own interests.

And a new threat had appeared on the horizon, at first faint, yet rapidly growing in significance. International terrorism was nothing new, but with the collapse of the Soviet Union, and resultant dispersal of much of its weapons technology and experts, the prospect of a nuclear weapon being acquired by a terrorist group became a real and present danger. While deterrence may be a major factor affecting the nuclear strategies of nation-states, it's of no concern to terrorists beholden only to their own ideologies rather than the survival of a country and its society. And while deterrence also requires the existence of some object of retaliation, terrorists offer no such convenient recourse, able to attack and then fade into invisibility.

Because of these and other issues, the dangers of nuclear weapons have evolved significantly since their invention in 1945, to the point where we must now consider questions that were irrelevant or inconceivable to the world of the Cold War. Whether we realize it consciously or not, much of our thinking on these questions is based on the perceptions, ideas, and depictions that arise from popular culture and media. I'll examine those in the next chapter to see how they have shaped our ideas, reflected our hopes and fears, and sometimes even directly influenced the world beyond our make-believe creations. As we will see, sometimes the most telling truths aren't found in the speeches of politicians, pronouncements and threats of generals, or technical analyses of scientists and strategists but instead in our imaginations.

THE ENDS OF THE WORLD: NUCLEAR WEAPONS AND POPULAR CULTURE

Before 1945, depictions of atomic bombs, nuclear warfare, and radioactivity, whether in the works of H. G. Wells or pulp science fiction or radio dramas, could be as fanciful and downright silly as their creators desired. Hiroshima and Nagasaki changed all that, imbuing atomic energy and weapons with a harsh reality. A great deal of misinformation and mythology would still persist, but from then on there would be some definitive parameters around it, engendering new images, tropes, fears, and hopes.

These would range from the personal, meaning the influence of the Bomb on individual human beings, to the broadest possible, meaning the end of the entire world and all of civilization. And their prevalence and influence would rise and fall with the general cultural zeitgeist regarding nuclear weapons—rising in times of great anxiety, and falling when the threat of nuclear catastrophe seemed to have faded. They would involve the full spectrum of media, including print, radio, television, motion pictures, visual arts, and even music.

Here, I'll concentrate mostly on a few representative films and television programs, since they're the most familiar, widely seen, and arguably most influential illustrations, but there are many more examples in the other media mentioned.

Alien Ultimatums and Radiation Mutants

The earliest postwar atomic tropes came almost exclusively from the pens of journalists who breathlessly described the atomic bombings, the now-public story of the Manhattan Project, and the where-do-we-go-from-here deliberations of statespeople and scientists. Chief among these was *New York Times* science writer William L. Laurence, who had been handpicked by General Groves to be the official amanuensis of the Manhattan Project in spring 1945 and the sole journalist made privy to most of its secrets. Laurence had been one of the few reporters who had been closely following atomic energy developments since the discovery of fission.[1] Laurence's Pulitzer-winning atomic reporting as the sole journalist at Trinity and over Nagasaki became a template for his colleagues who hadn't enjoyed his privileged position. Many of his ideas and turns of phrase rapidly became clichés, such as "mankind standing at a crossroads," "harnessing the power of the atom," and comparing atomic energy to "the philosopher's stone," among many others.

Beyond the newspapers and newsreels, it took a little time for the atomic bomb to make its way to motion pictures. Probably the first major film to directly address the Bomb was the 1947 *The Beginning or the End*, a rather staid docudrama that depicts with only approximate accuracy the story of the Manhattan Project and bombing of Hiroshima. The title itself is an example of one of Laurence's tropes: the notion that the Bomb and atomic power represent either the dawn of a new golden age of peace and plenty for all mankind, or heralds the end of civilization.

Except in radio serials, comic books, and somewhat more seriously, science fiction magazines and novels, no one was yet concerning themselves with trying to portray what an actual atomic war might entail. But two years after the US atomic monopoly was broken, the first major film to address the possibility of total nuclear devastation appeared: Robert Wise's 1951 *The Day the Earth Stood Still*.

As a lavish production from major movie studio 20th Century Fox featuring well-known Hollywood actors and top-notch technical talent, *Day* couldn't be dismissed by critics as just another cheap two-reeler Saturday afternoon serial. It was clearly a serious piece of work that tapped into all the prevalent anxieties of its time, not just

Figure 12 As the Cold War progressed, nuclear imagery began to pervade popular culture. *Atomic War!* comic, circa 1952. (National Archives)

the atomic bomb, but the flying saucer phenomenon and Cold War paranoia over Communists.

A saucer-shaped spacecraft touches down one summer day on the Mall in Washington, DC. Its alien passenger, Klaatu, and his towering robot, Gort, are immediately surrounded by the military, and Klaatu is shot by a nervous soldier. Spending a night at Walter Reed Hospital, where he's found by doctors to be perfectly human, he informs a government representative that "the future of your planet is at stake" and he has an urgent message to deliver to all of Earth's nations, but is told that "the evil forces that have produced the trouble in our world" (meaning, of course, the Communist bloc) make it impossible to arrange such a meeting. "Impatient with stupidity," as he informs the official, Klaatu decides he needs to get out among Earth's people to discover the basis for these "strange, unreasoning attitudes" and disappears from custody, soon befriending a young widow and her son while trying to make arrangements to deliver his message through the offices of an Einsteinian scientist, Professor Barnhardt. Eventually the authorities catch up to Klaatu and apparently kill him, but he's soon revived by Gort (in an overtly Christian metaphor) and manages to finally deliver his intended message to an international group of scientists assembled by Barnhardt: "The universe grows smaller every day, and the threat of aggression, by any group, anywhere, can no longer be tolerated. There must be security for all, or no one is secure. . . . We of the other planets have long accepted this principle. We have an organization for the mutual protection of all planets and for the complete elimination of aggression."

These words are strikingly similar to those heard during the earlier UN debates over the international control of atomic energy. Klaatu then proceeds to explain the means that his interplanetary alliance has devised to achieve these goals:

> The test of any such higher authority is, of course, the police force that supports it. For our policemen, we created a race of robots. [*He nods to Gort nearby.*] Their function is to patrol the planets . . . and preserve the peace. In matters of aggression, we have given them absolute power over us. This power cannot be revoked. At the first sign of violence, they act automatically against the aggressor. The penalty for provoking their action is too terrible to risk. The result is that we live in peace. . . . [W]e do not pretend to have achieved perfection, but we do have a system, and it works. . . . It is no concern of ours how you run your own planet. But if you threaten to extend your violence, this Earth of yours will be reduced to a burnt-out cinder. Your choice is simple: join us and live in peace, or pursue your present course and face obliteration.

Klaatu's farewell speech thus echoes what he had previously told Barnhardt: "I came here to warn you that by threatening danger, your planet faces danger." Although the concept of MAD had yet to be fully articulated by 1951, the system of "higher authority" that Klaatu's people have devised to protect them from themselves is essentially the same solution: peace or utter destruction. In Klaatu's civilization, the choice is imposed by all-powerful, impartial, emotionless robots; on the Earth of the Cold War, it's imposed by the doctrine of nuclear deterrence. (We will see a similar implacable logic employed to restrain the worst of humanity in several later films discussed below.)

As the first major film to present such apocalyptic themes, *Day* had a long-lasting influence not only on movie audiences but on political figures too. In 1985, recalling the movie, ex-actor and then president Reagan asked Soviet leader Gorbachev whether the USSR would help to defend the United States from alien invasion. (Gorbachev said it would.)[2] Later, Reagan remarked in a speech to the United Nations that "I occasionally think how quickly our differences worldwide would vanish if we were facing an alien threat from outside this world."

Another big Hollywood production, the 1954 *Them!*, sets out the second major narrative theme of the early atomic era—not atomic war itself, but the baleful consequences of the Bomb and radiation. Giant ants terrorize the New Mexican desert, killing people and wreaking havoc. Before the ant colony is found and destroyed, some of the winged breeding pairs escape to set up housekeeping on a ship and in the storm drains of Los Angeles, threatening the city. The creatures are all wiped out before they can fulfill the biblical prophecy quoted by one character, "The beasts shall reign over the earth." The monsters are found to be mutants created by radiation from the Trinity test. At the end of the film, star James Arness asks a scientist, "If these monsters got started as a result of the first atomic bomb in 1945, what about all the others that have been exploded since then?" Says the scientist, "Nobody knows. . . . [W]hen Man entered the atomic age, he opened a door into a new world. What we'll eventually find in that world, nobody can predict." Thus the coda of *Them!* neatly manages to work in several major motifs of the atomic age in a single statement.

What followed throughout the remainder of the 1950s were for the most part variations on these themes, generally done with far less intelligence and elegance as well as far more limited budgets. Either aliens came to Earth, whether to lecture us about the folly of our rush to atomic self-destruction or to take over (as in George Pal's 1953 update of Wells's *The War of the Worlds*, which features an ineffective H-bomb attack against the Martian invaders), or a human being or other living creature was horribly mutated, altered, or given mysterious powers by a bomb blast or some sort of radiation exposure. In the former, the Bomb exemplifies the ultimate weapon, the bringer of doomsday; in the latter, the atomic bomb or radiation is generally only a convenient plot device to create monsters, sometimes tinged with the implication that such horrors await us if we don't control the atom.

Some of the "radiation monster" movies did attempt to draw serious metaphoric parallels between their fictional monstrosities and the real terrors of nuclear war. Perhaps the most significant example came from the only nation to actually endure nuclear attacks, Japan. Coincidentally released at the end of 1954, the same year in which Japanese citizens had again fallen victim to the Bomb in the Castle Bravo fallout incident, *Gojira* (better known by its US title, *Godzilla*) was the first and best illustration of what became an entire film genre of giant monsters attacking Japanese cities. But unlike in those later films, the titular creature of the original movie is a terrifying, implacable engine of destruction, aroused from the undersea depths by H-bomb testing to represent the vengeance of nature itself on the human beings who would defile the earth with nuclear fire.

Such an interpretation was precisely what the filmmakers intended. For a people still recovering from the trauma of the atomic bombings, in a nation where public discussion of nuclear issues was discouraged, Gojira/Godzilla represented all the terror, shock, confusion, and overwhelming power of the weapons unleashed on them

Either aliens came to . . . lecture us about the folly of . . . atomic self-destruction . . . or a living creature was horribly mutated, altered, or given mysterious powers by a bomb blast.

less than a decade before. Although the 1956 reedited US version greatly tones down such themes (especially the monster's origins in US nuclear testing), scenes of destruction in the original Japanese film are eerily similar to the actual postattack footage of Hiroshima and Nagasaki. And in another historical parallel to the supposedly guilt-ridden atomic scientists of the Manhattan Project, the creature is ultimately stopped by a different superweapon called "the oxygen destroyer," the brainchild of a tormented scientist reluctant to use their own invention.

Few nuclear-focused US films of the period were so thematically and ethically complex, content instead to depict radiation as something that enlarges insects, arachnids, US soldiers, and spurned wives. As mentioned, some of these works paid lip service to the dangers of the atom, frequently with some concluding homily about "secrets Man was not meant to know," "choosing the right path between peace and doom," or similar clichés. None made any attempt to venture beyond their B-movie, drive-in boundaries to seriously consider the questions of nuclear war, except perhaps in the most oblique terms. In the 1956 film *This Island Earth*, several distinguished nuclear scientists are invited to join a secret group planning to "put an end to war." It develops that the group is actually led by refugees from outer space, and the war that they're looking to end is their own; their home planet faces extermination unless they can develop a defense. The scenes of the ruined and ultimately destroyed alien world along with the film's emphasis on both the destructive and constructive sides of atomic power are an echo of the inevitable culmination of an unbridled arms race.

Not With a Whimper . . .

As public fears shifted somewhat from the imminent possibility of global thermonuclear war to the more distant anxieties over long-term fallout effects with the rise of the test ban movement in the late 1950s and early 1960s, Hollywood finally began to get more serious about the atom. It also helped that the boom in low-budget, B-movie science fiction and horror films that had created such demand for radiation-spawned monsters was largely over by this time. And with atomic testing and peace demonstrations now solidly settled into respectable public discourse and learned political commentary, the atomic filmmaking baton could be passed from exploitation directors like Roger Corman and Bert I. Gordon to more respectable artists like Stanley Kramer, Sidney Lumet, and Stanley Kubrick. Kramer took up the baton first with his 1959 production of *On the Beach*, an adaptation of Nigel Shute's novel.

On the Beach is nothing if not a serious, almost self-consciously meaningful film—a status emphasized by its cast of A-list movie stars including Gregory Peck, Fred

Astaire, and Ava Gardner. There are no monsters, no aliens, and not even a single stock shot of an atomic explosion of the type that had become so boringly familiar in so many previous movies. In *On the Beach*, the nuclear war has already happened, and a collection of survivors fortunate enough to have escaped the initial carnage because they were living in Australia, or in the case of Peck's character, commanding a US Navy submarine, waits for the lethal shroud of fallout to drift down from the Northern Hemisphere and seal its doom. Each of the characters ultimately accepts and faces their inescapable deaths in their own way. No one remains at the end; Earth is dead.

Unlike even the previous major Hollywood Bomb-related films such as *The Day the Earth Stood Still*, Kramer's movie had an enormous worldwide impact, triggering passionate discussion and argument far beyond the usual circle of film critics and moviegoers. Politicians and public figures argued for its importance, such as Nobel-winning scientist and noted test ban advocate Linus Pauling, who reportedly claimed it might be "the movie that saved the world." Others condemned it for "defeatism" and remarked that its scenario of fallout literally killing the entire human race was scientifically preposterous. But whatever its scientific or artistic merits, *On the Beach* probably sparked more awareness of the ongoing presence and threat of nuclear weapons, more directly and across a broader audience, than any other nuclear-themed film of the era.

Debuting the same year as Kramer's film was a television series that ventured far beyond the kitchen-sink realism of studio-bound dramas to consider nuclear doomsday from far more metaphoric and fanciful angles in many of its episodes. For five seasons from 1959 through 1964, *The Twilight Zone* was creator Rod Serling's answer to the many creative restrictions and taboos forced on him in his earlier years as one of television's bright young men and leading writers. Some episodes focused on universal and eternal human themes, such as loneliness and alienation, and others on more immediate and contemporary subjects, including, naturally, nuclear war.

Possibly the best-known example is the early episode "Time Enough at Last," in which a self-centered, reading-obsessed bank clerk becomes the ultimate apocalyptic cliché: the last person left alive after atomic war. Taking his lunch break in the bank vault so he can read undisturbed, he peruses a newspaper with the prominent headline "H-Bomb Capable of Total Destruction" and then emerges to find the city in ruins. Another episode, "One More Pallbearer," spins a variation on this theme, with an eccentric millionaire staging an elaborately faked nuclear attack on New York City as a practical joke on some enemies from his past, only to find himself locked into a personal after-the-Bomb wasteland when his mind snaps. "The Old Man in the Cave" focuses on the fate of a ragtag group of survivors ten years after a nuclear war. And *The Twilight Zone* didn't concentrate solely on the postapocalypse world. In the episode "Third from the Sun," a family group led by an atomic scientist takes a desperate gamble to escape a nuclear holocaust only hours away. But probably the most realistic and raw *Twilight Zone* nuclear war episode was "The Shelter," in which close-knit neighbors in a typical US hometown facing a nuclear attack alert suddenly find themselves at each other's throats, clawing for entry into the only available fallout shelter in the neighborhood. Appearing in the midst of the early 1960s' programs for civil defense and fallout shelters, the story was a direct reflection of actual discussions among experts and pundits about the ethics of keeping intruders out of shelters built for only a single family.

By the years of Kennedy, the Cuban missile crisis, and talk of test bans and megadeaths, nuclear weapons were as familiar a subject and plot device in mass media as in newspapers and current events magazines, soon migrating from the more or less exclusive realm of science fiction to other genres, including the James Bond film series. But two motion pictures released in 1964 lifted filmic nuclear imagery and themes to new levels of influence and significance, to the point where both are still widely discussed more than half a century later.

Both films were based on popular novels—a US and a British one. Eugene Burdick and Harvey Wheeler's *Fail-Safe* first appeared in serialized form in the *Saturday*

Two motion pictures released in 1964 lifted filmic nuclear imagery and themes to new levels of influence and significance.

Evening Post, beginning just as the Cuban missile crisis was winding down at the end of October 1962. The novel thus perfectly struck a timely nerve with audiences that had just spent two weeks wondering if nuclear war was about to start, dealing with a technical mishap that sends a US bomber on a routine patrol beyond its usual fail-safe point to bomb Moscow.

The book quickly shot up the best seller lists and attracted the interest of Hollywood. Even before the film's production was completed, however, the producers were hit with a lawsuit from a rival film about accidental nuclear war: Kubrick's *Dr. Strangelove: Or, How I Learned to Stop Worrying and Love the Bomb*. Kubrick charged that the *Fail-Safe* novel and film had been derived from the same source as his own film: a 1958 British novel by Peter George (under the pen name Peter Bryant) titled *Two Hours to Doom* (*Red Alert* in the United States).

An autodidact genius with a dizzying range of interests, Kubrick had become fascinated some years earlier with the question of thermonuclear war and decided that it would be the subject of his next film. He immersed himself in research into the theories and practicalities of nuclear warfare, reading hundreds of books, articles, papers, and studies, and talking with dozens of experts. To provide the narrative basis for his film, he soon settled on George's novel, in which an unbalanced SAC general sends his bombers off to attack Russia without proper authorization. The book is suspenseful and deadly serious in tone, just like *Fail-Safe*.

As Kubrick and novelist Terry Southern began working on their screenplay, however, they found themselves increasingly struck by the fundamental absurdities of deterrence, nuclear war, and the idea of preserving peace by threatening doomsday. Inevitably, *Dr. Strangelove* evolved from straight drama into one of the greatest satires ever put on film. This shift in tone and Kubrick's refusal to espouse the accepted "Peace Is Our Profession" party line of SAC, as had several other Hollywood productions such as the 1955 Jimmy Stewart vehicle *Strategic Air Command* and 1963 *A Gathering of Eagles* starring Rock Hudson, cost Kubrick's production the official assistance and approval of the US Air Force. That didn't faze Kubrick, who was able to gather enough material from open sources to depict SAC policies, procedures, and equipment, including an extremely detailed B-52 bomber cabin, with an accuracy that both surprised as well as dismayed military and government officials.

The legal battle between *Fail-Safe* and *Dr. Strangelove* was settled out of court, with part of the settlement conditions allowing *Strangelove* to be released first, in January 1964. *Fail-Safe* was delayed until October, giving Kubrick the chance to get in the first strike in this particular cinematic Armageddon. Although *Dr. Strangelove* proved to be the bigger box office hit, both films were critically acclaimed and had a cultural impact extending well beyond the entertainment world. Considered together, they neatly encapsulate early 1960s nuclear anxieties, with *Fail-Safe* taking a starkly realistic stance and *Strangelove* a more absurdist, existential position. Viewed more than fifty years later, those different approaches to essentially the same subject matter and story have allowed *Dr. Strangelove* to retain much of its relevance and effect, while *Fail-Safe* seems more a relic of its time.

Kubrick's film not only acknowledges the ultimate absurdity of nuclear warfare but positively revels in it, as is evidenced immediately by the names of the characters. The deranged general who sends his B-52s to attack Russia is Jack D. Ripper, for example, befitting a man about to cause the greatest act of mass murder in history. The title character, a wheelchair-bound ex-Nazi strategist who speaks of megadeaths and doomsday weapons with casual coldness, is an amalgam of Henry Kissinger, Edward Teller, and then fashionable nuclear war intellectual Herman Kahn. The crisis unfolds across only three main locales: Ripper's Burpelson Air Force Base, the war room of the Pentagon, and the cabin of the renegade B-52 that threatens to trigger a Soviet doomsday machine that will shroud the entire earth in "radioactive cobalt thorium-G," killing all life aboveground.

Strangelove offers the prospect of a select few "top government and military men" surviving in underground shelters, where "animals could be bred and slaughtered" for food, and "highly stimulating" females will be provided in order to quickly repopulate the United States. Before the worst happens and the Soviet doomsday machine is indeed activated at the end of the film (depicted by a stock-footage series of H-bomb explosions to the tune of the sentimental World War II ballad "We'll Meet Again"), Kubrick manages to hit every touchstone and pet phrase of contemporary nuclear strategy, from generals speaking of twenty million dead Americans as little more than getting our "hair mussed," to various Freudian allusions surrounding nuclear weapons and potency, to moralistic debates about the ethics of first strikes or how to ensure that the war can be "won."

In *Dr. Strangelove*, Kubrick boils down the essence of nuclear war to its most primal human elements: sex, animal cooperation, and physical survival. He focuses less on the technology than on the human beings who have constructed a shaky system to achieve an impossible sense of security and thus entrapped themselves within the mercy of that system's ineluctable logic when its inadequate safeguards collapse. If the details of SAC's policy of deploying its planes and keeping them on alert (and indeed, SAC itself) have changed or been abolished since *Strangelove* was produced, these basic concerns of humanity have not, and are still enshrined in US nuclear weapons policies and weapons systems.

Whereas *Dr. Strangelove* didn't hesitate to make human beings look ridiculous, *Fail-Safe* clearly strives to be another important motion picture of our time in the vein of its predecessor *On the Beach*. Director Lumet's straightforward and staid mise-en-scène approximates documentary, with no frivolity such as incidental music, humor, or romantic subplot permitted to intrude on the proceedings. Like Kubrick, Lumet confines most of the action to a handful of settings, including the cockpit of a SAC "Vindicator" bomber, an underground White House bunker where the president confers with the Soviet premier via a hotline, the war room of SAC headquarters where the progress of the errant aircraft is graphically displayed on a huge electronic screen, and a Pentagon conference room where generals and advisers ponder the situation and pass along advice to the president.

Fail-Safe makes it clear that the crisis was directly caused by a technical flaw that issued false attack orders to the bombers waiting at their "fail-safe" points, but suggests that humans have become too dependent on such technical wonders to control their fates and therefore share in the responsibility for what could happen. (*Strangelove*, on the other hand, doesn't let humanity off the hook with any such convenient excuse.) By the film's conclusion, the merciless logic of deterrence forces the US president to make a horrific choice in order to avoid total nuclear war, but unlike in *Strangelove*, the world is spared Armageddon.

In both films, the crises that threaten to end human civilization all occur within the sight, awareness, or control of only a select few military men and civilians, with the rest of the world's population completely oblivious to what's going on. This, of course, was in an era long before twenty-four-hour news networks and the internet, so it would have been quite possible that most of the citizens of the United States and USSR wouldn't know anything untoward was happening until H-bombs began falling on their cities. But narratively speaking, it allows both films a certain claustrophobic viewpoint that actually ratchets up the tension. Intentionally or not, it also emphasizes the utter helplessness of the public in the face of thermonuclear war and ultimate responsibility of the leaders who hold the fates of entire populations in their hands. As we will see, later nuclear war films had to incorporate public reactions to their crises to preserve a convincing verisimilitude.

One minor, if notable, film from this period, however, did explore nuclear war from the perspective of the average American: 1962's *Panic in Year Zero*. Though a low-budget offering from a noted exploitation independent production company, it offered a lurid, less sophisticated, yet nonetheless more direct take on atomic war as might be experienced not at the exalted, abstract remove of senior leaders but rather at ground level by regular people. In the film, a typical American family embarks on a camping trip early one morning in their station wagon and trailer, only to see a bright flash in the sky and hear an emergency broadcast over the car radio.[3] The scene in which the puzzled family pulls over on the side of the road at an overlook to gaze at the enormous mushroom cloud rising over the Los Angeles they've just left is still chilling, with the obvious low-budget special effect giving it an aura of unreality that somehow makes it more effective.

Realizing that they have enough supplies along to survive on their own for the moment at least, the family decides to continue on to their campsite in the hills. They begin to encounter signs of the panic and confusion gripping the populace, such as stores being stripped bare of food, merchants gouging customers for essential supplies, and terrified people resorting to violence as they desperately flee the impending prospect of more H-bomb attacks. The family find themselves caught in a primal struggle for survival, eventually forced to hide in a cavern, and to kill or be killed. Finally, the end of the brief war and restoration of civil order (under martial law, at least) is announced on the radio, and the family starts back to whatever remains of civilization. The United Nations proclaims that the world will now start again at "year zero" (hence the title), and the film closes with the words "There Must Be No End—Only a New Beginning."

As a work of cinema, *Panic in Year Zero!* is hardly comparable to the work of Kubrick or Lumet. But as an example of the nuclear paranoia of everyday American citizens and how they might react in the face of atomic war, it's a telling document. In one of the many civil defense films of the 1950s, titled *You Can Beat the A-Bomb*, a confident, supremely competent father prepares his family for a nuclear attack and then tells them, "Nothing to do now but wait for orders from the authorities and relax." In *Panic*, though, despite his reassurances that rightful law and order will eventually be restored, the father advises his family that until then, they're completely on their own. There will be no depending on the "authorities" to save anyone; for the time being, the law of the jungle will prevail—the same law that, as *The Twilight Zone* has shown, allows you to keep neighbors out of your fallout shelter in order to save your own family.

As the 1960s progressed, fears of nuclear apocalypse receded in American consciousness and consequently popular culture, supplanted by other preoccupations such as the Vietnam War, civil rights, and the schism between youth culture and the "Establishment." Nuclear war was still a convenient narrative device when needed to end the world, as in the 1968 film *Planet of the Apes* or a number of lesser science fiction works. But any attempts to directly address nuclear warfare with the seriousness of *Fail-Safe* and *On the Beach*, or even the irreverent satire of *Dr. Strangelove*, were largely absent.

Two notable exceptions, however, came with Peter Watkins's 1966 production *The War Game* and Joseph Sargent's

1970 movie *Colossus: The Forbin Project*. The former film was perhaps the first serious attempt to actually portray a nuclear attack as it might happen, with a documentary accuracy. Produced for the BBC and intended for television broadcast with a length of less than an hour, it turned out to be far more than the BBC had expected: graphic, unyielding, and visceral even by today's standards. Watkins offers no comforting homilies or reassuring images, only the unrelenting horror of English citizens being burned and blasted, dying of radiation sickness, being shot for looting, and seeing their families, neighborhoods, and entire society destroyed around them. The BBC promptly banned the film from British television, deeming it "too horrifying for the medium of broadcasting," and it was relegated to special screenings and film festivals until finally being broadcast in 1985.

The US movie *Colossus: The Forbin Project* was hardly as graphic and was a conventional Hollywood feature film, yet it contained an even higher (implied) body count than *The War Game*. A massive supercomputer named Colossus, sheltered inside an impenetrable redoubt in the Rocky Mountains, is placed in control of all US defenses, with its creator, Dr. Charles Forbin, promising that its dispassionate nature will prevent any miscalculations or intemperate actions that could cause nuclear war. But the self-aware Colossus, with access to all world communications and intelligence, soon discovers an identical Soviet counterpart named Guardian. When its human interlocutors refuse to connect Colossus with Guardian, the system launches an ICBM at the USSR, and Guardian promptly retaliates, with both computers doing precisely what they were intended to do. The humans acquiesce in time for Colossus to agree to intercept and fend off Guardian's attack, though not fast enough to prevent Colossus's missile from taking out a Russian oil complex. Still, peace is preserved for the moment. Once more, the majority of humanity has faced down extinction without even knowing it.

But Colossus has other ideas, joining with its Soviet opposite number to form a united, world-girdling artificial intelligence in complete control of the earth's entire nuclear arsenal. Colossus proclaims a new millennium of peace, announcing to the world, "The object in constructing me was to prevent war. This object is attained. I will not permit war. . . . I will restrain man." To demonstrate its power, Colossus detonates a US ICBM within its silo, thwarting a joint US-Russian human attempt to covertly disarm their missile warheads.

If the theory of deterrence has its own peculiar logic, then that logic is completely manifested in its purest form within the silicon heart of Colossus. Generally remembered as an exploration of the time-honored science fiction trope of the omniscient computer and dangers of artificial intelligence, the film is a classic Frankensteinian fable about the human arrogance of "playing God"

(Forbin even muses at one point that "Frankenstein should be required reading for all scientists"). But it's also a metaphoric depiction of the perils of placing too much confidence in automatic decision-making systems, whether mechanical ones like Colossus or abstract ones like deterrence theory, to protect us from ourselves. If we choose to delegate responsibility for our own survival to some outside entity or agency, and thus forgo the moral choices that make us human, then we will have to submit to the remorseless outcomes that result from that choice, like it or not. Or as Colossus observes, "Freedom is an illusion."

Armageddon in the Backyard

With the 1970s era of East-West détente and arms control agreements, popular culture continued to mostly forgo any direct engagement with nuclear war except its convenience as a plot device. Yet as public fears sharply intensified in the following decade with renewed nuclear saber-rattling between the United States and USSR, so did cultural attention to the end of the world. The 1980s brought the most influential nuclear war film since *On the Beach*, a work that even Reagan cited as having a profound effect on his own views toward nuclear war and weapons: the 1983 television movie *The Day After*.

Following in the footsteps of *The War Game*, but with far more ambition and a broader scope (as well as a far greater budget), *The Day After* was the first major US film that attempted to accurately depict the effects of nuclear war on average civilians. Director Nicholas Meyer, a man on a mission, doggedly fought off network attempts to censor some of the more graphic scenes of the film, including an extended attack sequence on Kansas City and nearby communities showing people being vaporized, incinerated, and killed in various ways by the exploding warheads along with the havoc they unleash.

The attacks on the United States are preceded by a period of increasing international tensions, mirroring much of the actual world situation at the time, until negotiations collapse and the world slides into war with terrifying swiftness. Meyer follows a representative group of people, including a doctor, a farmer and his family, college students and professors, and SAC missile staff stationed at nearby ICBM bases. Even more terrifying than the actual attack sequences may be the shots of Minuteman missiles being launched from silos in the Missouri farmland as the locals helplessly gaze at the missiles trailing white contrails into the sky on their way to Russia, knowing that inescapable doomsday is only minutes away.

Heavily promoted as a major television event for months before its November 20, 1983, broadcast on the ABC network, *The Day After* was a ratings smash, both terrifying viewers with images that would still haunt them many years later while familiarizing them with basic facts about nuclear warfare that many people had never before realized. Younger viewers who hadn't been around in the earlier Cold War days of civil defense drills, fallout demonstrations, and missile crises were unfamiliar with matters such as counterforce versus countervalue targeting, radiation sickness, electromagnetic pulses, or the fact that once fired, ICBMs can't be recalled or defended against. After the broadcast, ABC and other television networks, along with local community and national groups, held discussions and debates on deterrence, disarmament, and nuclear winter.

As noted above, the film's impact reached directly into the White House. President Reagan and some advisers saw it in a private screening before the public broadcast and were deeply affected, with Reagan confessing to his diary that the movie left him "greatly depressed." Other officials expressed similar sentiments. Reagan later wrote in his memoirs that *The Day After* was one of the major factors leading to the 1987 Intermediate-Range Nuclear Forces Treaty with the Soviet Union.

Some critics praised the film lavishly, while others condemned it as too sensationalistic or politically motivated. The general US audience, though, affected on a deeply visceral level, seemed to disregard such critiques. They had just watched thousands of their fellow Americans being wiped off the earth by incomprehensible forces, and thousands more dealing with the utterly bleak aftermath of struggling to survive in a blasted, radioactive wasteland with almost every familiar touchstone of civilization destroyed. Those visions transcended politics and hit on a fundamental level. Experts might have quibbled over the technical and narrative details of the film—most of them arguing that in fact, the reality of nuclear war would be far worse than portrayed—but *The Day After* seemed to have enough of an effect, and one far more powerful than *On the Beach*, *Fail-Safe*, *Dr. Strangelove*, or any previous works. Whereas *On the Beach* was stately and almost melodramatic, quite free of any graphic images, *The Day After* brought nuclear annihilation home to everyone's neighborhood and family. Whereas *Fail-Safe* and *Dr. Strangelove* dealt with nuclear war with a certain lofty abstraction, showing the people who would launch the weapons, *The Day After* captured what happened to the people on whom the weapons were unleashed.[4]

The year following *The Day After* saw the release of a new British take on nuclear war, the BBC drama *Threads*. Along with *The War Game* and *The Day After*, this production caps an unofficial filmic nuclear war troika and is perhaps the most uncompromisingly graphic of the three, not only in its depiction of a nuclear attack on Britain, but in the lingering aftermath of a global nuclear winter, long-term effects of radiation, and collapse of human society.

Whatever nuclear nightmares *The Day After* failed to give British audiences were more than amply supplied by *Threads*.

The imagery of the atom, from the ubiquitous outline of the mushroom cloud and the nucleus surrounded by spinning electrons, to the most graphic portrayals of death and destruction, has been intertwined with the realities of nuclear weapons ever since the first newspaper reports of Hiroshima and Nagasaki. They've evolved in parallel with those realities since then, sometimes distorting the truth through exaggeration or understatement. But because narratives and images have such powerful effects on the human psyche and perceptions, whether consciously acknowledged or not, they influence our actions. Our beliefs, our fears and hopes, and our personal knowledge or lack thereof can affect what political platforms and policies we support, the candidates we elect, the demonstrations we attend, the rumors we spread, or the truths we defend. In the nuclear realm, this can mean everything from the adoption or rejection of arms treaties and defense budgets for national leaders to the decision to build a home fallout shelter or contribute to a disarmament campaign.

But beyond these examples, the attitudes and ideas stimulated in us through cultural imagery and social expression influence our ultimate mindsets—whether toward hope and optimism along with a constructive resolve to avoid nuclear war, or despair, pessimism, and helpless

The imagery of the atom . . . has been intertwined with the realities of nuclear weapons ever since the first newspaper reports of Hiroshima and Nagasaki.

resignation to inevitable oblivion. That's why it's important to examine the images and stories on which much of our understanding about nuclear weapons is based. Ultimately, that understanding will determine how we will continue to live with them in the future, and for how long—or whether it might be possible to eliminate the threat of nuclear weapons for all time.

THE CHOICE

When the Cold War passed into history in the early 1990s, nuclear weapons did likewise in the minds of many people. Thoughts of thermonuclear war, deterrence, mutual assured destruction, nuclear winter, ICBMs and bombers, civil defense, and the living envying the dead—all of it was relegated to the past as something about which we no longer had to worry.

If they stopped to think about it, some folks might have realized that all those thousands of nukes didn't simply vanish when our former adversary, the Soviet Union, collapsed. The United States still had its missiles and submarines and bombers after all. But of course, *we* weren't going to attack anybody, and with our mortal enemy gone, who was left to threaten us? No, the worst was over now, the Damoclean sword of nuclear oblivion no longer hanging above humanity's head, and the United States could now concentrate on beating its nuclear swords into plow-shares, spending all the money sunk into bloated defense budgets into more positive social enterprises. The country might keep a few nukes around as insurance to deter China or some other nation that might still decide to acquire the Bomb, but for all practical purposes, the nuclear threat was basically over.

Would that it had been so. Before the end of the century, some would find themselves longing, if perhaps reluctantly, for the convenient simplicity and relative stability of the bipolar world of the Cold War, when all the United States had to worry about was the USSR rather than loose nukes, rogue nuclear powers, and terrorists. It's true that there are now far fewer nuclear weapons in the world than there were in, for example, 1983. Still, there are more than enough to destroy the current state of human civilization if they were used.

And while the prospect of a massive thermonuclear war involving thousands of weapons unleashed all at once is unlikely, barring some kind of human error, madness, or technical glitch that lights the spark, even a small conflict between two members of the nuclear club could have profound effects. A 2019 paper published in *Science* examined a hypothetical nuclear war between India and Pakistan using several hundred weapons. The authors concluded that the smoke and soot released as well as their spread around the globe from such a conflict would reduce sunlight by

20 to 25 percent, cooling temperatures by 2° to 5°C, and reduce precipitation by as much as 30 percent, with these effects lasting more than ten years. Because Indian and Pakistani cities are so densely populated, they note, "even a war with 15-kt weapons could lead to fatalities approximately equal to those worldwide in WWII and a war with 100-kt weapons could directly kill about 2.5 times as many as died worldwide in WWII." And all this in only a week.[1]

Facing these new uncertainties, new questions naturally arise. What can or should we do now? Has deterrence actually worked all this time to prevent nuclear war or is there another explanation? Either way, is there any reason to maintain nuclear arsenals in a world where deterrence is basically obsolete? Do nuclear weapons have any actual usefulness otherwise? And is it really possible to get rid of them for good?

Why Are We Still Here?

Along with a widespread sense of triumph among many that we had somehow "won" the Cold War, the breakup of the Soviet Union and the accompanying collapse of the global bipolar power balance raised the question of how it happened that despite all the nuclear crises and bomb tests and unrestrained weapons building, the world had somehow managed to survive all those decades since

Nagasaki without a nuclear weapon being used in anger. Some pundits claimed the answer was obvious, that deterrence had worked, and thus all the expense and arms racing and threat hurling of the Cold War had been worth it. It had been a rough and stressful time, no doubt, but the wisdom of constantly striving for strategic dominance and never letting your opponent pull too far ahead had been demonstrated. Every one of those crises and confrontations and near misses had ultimately been resolved by deterrence.

Others contested that view, arguing that even if deterrence had perhaps helped, we had only avoided nuclear conflict through sheer luck. Any of the many nerve-tingling dustups of the Cold War could have easily gone the other way toward disaster, and many examples could be cited where only random chance prevented it. What if Kennedy had followed his initial impulses as well as the strong recommendations of all his advisers and tried to destroy the Cuban missile sites right after discovering them? Or if the officers on duty during any of the many false alarms from early warning radar networks on both sides had concluded that yes, their country *was* under attack and immediate retaliation was required? The revelation of formerly classified information gave such contentions some sharp teeth. So maybe we had only been lucky.

And still more answers were possible. Maybe it all came down to some ineffable quality of the human soul, something that stays the hand of those with the power to press the nuclear button when their finger is poised over it in the last, decisive moment. Maybe humanity has evolved and learned the error of its ways. Maybe, ever since Nagasaki and the end of World War II, we were living in a "Long Peace," as historian John Lewis Gaddis and others characterized it, in which global conflict between major international powers had become a thing of the past, whether or not nuclear weapons had brought it about.[2] It was even possible to synthesize all the above in a sort of amalgam of an explanation, holding that some kind of unspoken yet powerful nuclear taboo had arisen. No matter the circumstances, in the final analysis no one wanted to cross the line and be the first to again use a nuclear weapon in war, because the world beyond was only darkness and terra incognita.

While some maintained that the question was irrelevant, more perceptive thinkers realized that there was an important reason to try to find an answer. Because whatever that answer might be, whether sheer luck or deterrence theory or the better angels of our nature, it might not last forever.

Wild Cards

Aside from a lot of expansive theorizing from various theorists and learned analyses from various analysts, the end of the Cold War brought new and serious problems. As noted in chapter 6, the loose nukes concern was largely resolved by the Nunn-Lugar Cooperative Threat Reduction program, but larger and more intractable issues loomed.

One lesson that the years of living under the umbrella, or shadow, of nuclear superpower domination had driven home to the smaller, less powerful nations of the world was that the possession of the Bomb imbues international prestige and influence that even the big players can't afford to ignore. The leaders of Third World nations such as Pakistan or Libya realized that there was no conceivable way for them to plausibly challenge the United States or Britain or Russia on a military basis, but that didn't matter. Merely having the capability to nuke a US city or Russian army base would be enough to make those countries pay attention to them. MAD, these leaders understood, didn't have to be complete. Just killing a few thousand people at one stroke was enough because none of the great powers would be willing to risk paying such a price. It was a way of bending deterrence to new ends in the pursuit of international influence.

With this in mind, nuclear ambitions long restrained by the United States and USSR were now unshackled, and proliferation became a growth industry, helped along by former Soviet scientists and engineers whose need for money overcame any ethical scruples, and opportunistic independent operators such as the Pakistani metallur-gist A. Q. Khan, who set up a profitable operation selling uranium enrichment technology and weapon designs to Libya and North Korea while also helping his own coun-try build its bomb. These activities weren't ignored by the established members of the nuclear club, even those who were unacknowledged, such as Israel, which had in 1981 wasted no time in destroying an Iraqi reactor believed capable of creating bomb-grade plutonium just before it became operational.

Lurking in the background of all these threats during this time, and not truly coming into its own as a major is-sue until the September 11, 2001, attacks on the United States, was the question of terrorist nuclear weapons. The 9/11 attacks may have killed around three thousand peo-ple, but officials clearly recognized that such a toll would be miniscule if the Al-Qaeda conspirators had used a nu-clear weapon rather than three fuel-laden airliners on that fateful day.

As atomic scientists had pointed out at the dawn of the nuclear age, there's no such thing as the secret of the atomic bomb. The basic principles governing the workings of atomic and thermonuclear weapons are open-source information, arising from the fundamental physics of the universe. And while the specific technical details of bomb design can be held close to the vest, the basic principles of implosion, gun-type weapons, and the variations thereof have been open knowledge for decades. A Princeton college student caused a major uproar in 1976 when he designed (but did not actually build) an atomic bomb as a school project, using publicly available material. Now, far more detailed information is available with a simple Google search.

Fortunately, however, even if the knowledge of how to build a nuclear weapon is widely available, far more is still required to accomplish the task. The biggest hurdle, and thus the best chance to prevent it happening, is acquiring the enriched uranium or plutonium fuel. Not much of either substance is needed: either about 20 to 110 pounds (9 to 50 kilograms) of highly enriched uranium, or about 4.5 to 14 pounds (2 to 6 kilograms) of plutonium, depending on the bomb design. The hitch facing would-be homemade bomb builders is that both are perhaps the most closely guarded resources on the planet.

That fact hasn't discouraged terrorists or rogue states looking to build a weapon; they are in general are among the most motivated—and patient—individuals in the world. It's not necessary to collect all the material for a bomb at once; it can be done over a period of time, a little here, a little more someplace else, to avoid attracting attention. The direct approach is possible by breaking into a facility containing the stuff, and in fact this has been attempted on more than one occasion. But bribery or blackmail are more subtle and less risky, and again, have been tried.

Another hurdle is that even with sufficient nuclear fuel to construct a weapon, it's an extremely complex undertaking requiring great precision in creating, machining, and assembling the various components, and all must be done within strict tolerances to ensure that the device produces a nuclear detonation when it's triggered. Even if the bomb makers or experts working for them possess the requisite knowledge and ability, a terrorist organization or relatively poor nation might not have the basic technical resources to succeed. This does not in any way eliminate the danger or imply that it can be disregarded but instead simply emphasizes that the nuclear terrorism scenario is more complicated than is sometimes portrayed.

Terrorists have other nuclear options. A rogue nation could either ally itself with or exploit a terrorist group to nuke a city, footing the bill and providing everything else required, using the terrorists to conceal its own responsibility and avoid reprisal. Multiple devices could be covertly placed in several cities in advance, followed by an ultimatum. Or as we've seen in chapter 5, there's the cheaper and easier option of a dirty bomb. Such possibilities keep intelligence professionals and military leaders up at night.

The emergence of nuclear terrorism and uncontrolled proliferation to other powers as new threats in the twenty-first century exist alongside some of the perennial threats

The nuclear terrorism scenario is more complicated than is sometimes portrayed.

of the past. US and Russian nuclear forces may have reduced in size, but many of them are still maintained on a hair-trigger alert status as if the Cold War had never ended, ready to be launched in response to another false alarm or technical glitch. In some ways, the increased sophistication and pervasiveness of our detection and warning systems, featuring satellite networks and other technology that was only a dream in the past, actually increases, not decreases, the risks of catastrophic accidents because of the added layers of complexity coupled with greater speed that place them beyond the comprehension or direct control of human beings.

And of course, as long as the weapons themselves exist, so will the option of somebody deciding to use them, not out of madness or by mistake, but by a conscious, reasoned decision. Which raises another question: Do nuclear weapons really have any practical use after all? Is the reason none have been used for their intended purposes since Nagasaki simply because no one has really managed to find a purpose for them, aside from preventing others from using them? If we've yet to find any use for nukes after almost seventy years, is it possible that they're simply obsolete, and if so, why are we still keeping them around? Can't we simply just collectively decide that they're not worth the risk and expense, and get rid of them forever?

If we've yet to find any use for nukes after almost seventy years, is it possible that they're simply obsolete, and if so, why are we still keeping them around?

The Right Tool for the Right Job

Even before Hiroshima, some argued that there was little military point in using the Bomb on Japan, a nation already on the verge of collapse. Comparable destruction to what was expected from the Bomb had already been visited on Tokyo and almost every other major Japanese metropolis by conventional explosives as well as firebombing. Razing a city to the ground with a single bomb and single airplane instead of hundreds at one blow might be more efficient from a bloody perspective, but not qualitatively different. The main impact would be psychological.

After World War II, and even for some time after the Russians got the Bomb, there were two major schools of thought. One considered the atomic bomb a weapon of genocide, utter destruction, and the end of the world; the other side thought of it as merely another tool in the military toolbox, just bigger and noisier. In the governmental and military circles that counted, the latter view prevailed, and so some leaders seriously considered using atomic bombs in Korea or other confrontations. The advent of the hydrogen bomb and its vastly increased destructive power made it somewhat harder to justify nuclear weapons as another handy military tool, but many still claimed that smaller weapons could be employed for limited tactical purposes without setting off World War III.

And yet when opportunities arose to demonstrate the alleged utility of nuclear weapons, no one took that step. They threatened, they made preparations, and they sent their bombers to poke at their enemy's borders and placed their missiles on alert, but they always backed off at the last possible moment.

Why? Could it be that those who had to make the decision—always a single person, either a president or premier—realized that the certainty of what nuclear weapons would do was outweighed by the awful uncertainty of what might follow their use? That there still existed other options or actions that could be taken? That when considered in terms of achieving any sort of practical, rational objective, nuclear weapons were actually essentially useless and therefore not worth it? That nuclear taboo or not, no world leader wanted to be the one to break it after all this time and perhaps become, in the words of *Dr. Strangelove*'s President Merkin Muffley, "the greatest mass murderer since Adolf Hitler"? Any and all are possible, and all that matters is that everyone faced with that choice since World War II has made the right decision—so far.

When the hydrogen bomb was created, its big selling point was its vast destructive power, far greater than the "mere" atomic bomb. The belief was that bigger bombs, greater yields, were better by definition, so that, for example, a twenty-megaton H-bomb was obviously a superior weapon to a twenty-kiloton atomic bomb. Such notions were fostered by the fact that the ultimate yield of fission weapons is limited by physics, while the yield of hydrogen weapons can be increased indefinitely. So for a time, both the United States and USSR focused on higher and higher yields, apparently hoping to intimidate the other into submission with ever bigger and better bombs.

But both realized that there's a point at which such a pursuit becomes counterproductive. There are only so many enemy targets, and they're only so big. One of Oppenheimer's original objections to the hydrogen bomb was that it would only be useful for really big targets— that is, large urban areas—and the USSR simply didn't have many of those; in fact, the United States contained better H-bomb target candidates. His assertions were ignored, but later on, strategists began to see that he had a point, and started to talk about concepts such as "overkill" and "making the rubble bounce." The Soviet Union essentially ended the argument when it designed a hundred-megaton bomb in 1961 and then realized that it couldn't test the weapon because it would cause too much damage on its own country. Scientists instead settled for testing a fifty-megaton design, which caused enough havoc on its own.

Another factor in reversing the trend toward bigger yields was that the ever-increasing accuracy of delivery systems, specifically ICBMs, made it unnecessary. A missile that can hit within just a few miles or yards of its target doesn't need to destroy such a broad area as one with far less accuracy.

Now, in the late twentieth and early twenty-first centuries, guidance and targeting technology have achieved enormous sophistication, with smart bombs, laser-guided munitions, and remotely piloted and autonomous drones that can zero in on a single building in the midst of a large city. If the objective is to take out a particular military headquarters or command center, it's not necessary to blanket the entire area with destruction. Countries can focus precisely on what they want to destroy. So what's the military utility of a weapon that levels hundreds of square miles at a blow? Former nuclear weapons researcher and Defense Threat Reduction Agency director Stephen M. Younger notes,

> Military planners are finding that explosive force is no longer the sole arbiter of military victory—a small amount of force applied at the right time and the right place can sometimes achieve the same result as a nuclear weapon. . . . Today, when confronted with the immediate use of weapons of mass destruction against our cities or military forces, our only options are to do nothing or launching a preemptive nuclear attack. New technology offers better solutions.[3]

In the beginning of this book, I presented the idea that nuclear weapons are fundamentally different than conventional weapons and they're much more than simply very big bombs. But there's a certain contradiction here as well. Many of the factors that appear to have prevented their use for more than seventy years are based on their great power, and controlling that power after it's been unleashed might be impossible. In other words, no one's used nuclear weapons *because* they're just very big bombs—so big that they're basically useless in any practical sense.

Their main utility is not practical but instead psychological. We see them as harbingers of doomsday and bringers of oblivion, which makes them tools with which to threaten and coerce each other while also making us hesitant to actually use them for their intended purpose. The director of the Rethinking Nuclear Weapons project, Ward Wilson, makes a similar point in his *Five Myths about Nuclear Weapons*:

> Nuclear weapons are implements that we manage and use as we wish. If someone said, pointing to a hammer on the workbench, "That hammer is beyond our control," we would think that person was pretty peculiar. Why are we inclined to view nuclear weapons differently? . . . Much of the hold that nuclear weapons have on us is *psychological*. Their

size in our mind's eye is not related to their size in the real world of pragmatic consequences. They are wrapped in a shroud of sixty years of rhetoric and hyperbole.[4]

This also provides us with an opportunity, though, if we choose to take it. As President Kennedy said in what became known as his "Peace Speech," one of the last major speeches of his life, "Our problems are manmade; therefore, they can be solved by man. . . . [N]o problem of human destiny is beyond human beings."[5] Can the problem of nuclear weapons be solved?

The Zero Option, More or Less

The concept of completely eliminating nuclear weapons from the face of the earth is hardly a new idea. It was there at the beginning, contained in the first efforts to reach international agreement in the wake of Hiroshima and Nagasaki, and has returned regularly ever since, sometimes voiced by peace activists, and sometimes by political leaders. It's one of those notions that almost everyone supports in principle, whether or not they think it's actually possible.

The chief objection always boils down to the stark truth that the Bomb can't be uninvented. Even if every nuclear weapon in the world were dismantled and destroyed, the knowledge to build them can't be expunged from human consciousness. All it would take is for one nation or one terrorist group to build another one, and we'd be right back where we started.

It's difficult to argue against this point. Even former secretary of defense William Perry, a staunch disarmament advocate, observes that "we can't repeal $E = mc^2$."[6] Once a technology exists in the world, it may become obsolete or fall into disfavor, but it continues to exist in some form, even if only in the forgotten texts of the past. A technology as powerful and seductively attractive to anyone seeking power as nuclear weaponry is unlikely to be forgotten.

But if a technology cannot be forgotten, it can still be controlled. If all its tools and implements and devices cannot be eliminated, they can still be reduced in numbers and importance. With a technology that is simultaneously as useless and dangerous as nuclear weapons, there's an overwhelming incentive to strive for a goal of complete abolition, even if it's ultimately unattainable in any practical sense.

In a widely influential editorial published in the *Wall Street Journal* on January 4, 2007, Perry joined with fellow senior statesmen George Shultz, Henry Kissinger, and Sam Nunn in calling for "A World Free of Nuclear Weapons." The four men combined their decades of expertise and experience in dealing with nuclear weapons to argue for a major new effort by the United States to move the world toward that seemingly impossible goal, warning that "unless urgent new actions are taken, the U.S. soon will be compelled to enter a new nuclear era that will be more precarious, psychologically disorienting, and economically even more costly than was Cold War deterrence. . . . Will new nuclear nations and the world be as fortunate in the next 50 years as we were during the Cold War?"[7]

For several years following that editorial, there seemed to be hopeful signs of, as Perry and his colleagues called it, a "reassertion of the vision of a world free of nuclear weapons" with "practical measures toward achieving that goal."[8] Unfortunately, that vision has clouded and faded considerably since then. At a public discussion on nuclear weapons held in Santa Fe, New Mexico, in December 2016, Perry told the audience, "I'm sorry to report to you that the likelihood of a nuclear catastrophe today is greater than it was at any time since the Cuban Missile Crisis."[9]

That was just before a new presidential administration took office. It proceeded to unilaterally withdraw from several major disarmament treaties and agreements—including the 1987 Intermediate Nuclear Forces treaty and the Joint Comprehensive Plan of Action with Iran (commonly known as the "Iran Deal"), which had, if only temporarily, suspended Iranian nuclear activities—and generally increased international tensions with nuclear powers including Russia, China, and North Korea. With again-expanding defense budgets and plans for new nuclear weapons development by both the United States and Russia, the world seems to be moving in the opposite direction from any implementation of the proposals outlined by the 2007 editorial, and is becoming less secure as a result. The only US-Russian arms treaty still in effect, New START, will automatically expire in 2021 unless renewed. The 1996 Comprehensive Test Ban Treaty still awaits ratification before going into effect. Shortsighted, politically partisan opposition to such treaties ignores the fact that they do more than simply limit or abolish weapons. They establish vital transparency and verification measures that increase, not diminish, security for all involved parties—security that's lost when the treaties are ripped up.

Some positive developments have emerged during the past several years. In January 2021, a new US presidential administration took office, promising a more conscientious approach to international relations and nuclear arms control. Earlier, in July 2017, the first legally binding international agreement to limit and ultimately abolish nuclear weapons, the Treaty on the Prohibition of Nuclear Weapons, was adopted in the UN General Assembly and signed by eighty-six member states, and went into effect in January 2021. Unfortunately, none of the signatory nations are actually nuclear-armed states, making the treaty an important yet essentially symbolic statement—for now.

Still, we have faced dangerous times before and survived them. Recently, in 2020, humanity confronted and, despite many setbacks, missteps, and terrible losses, ultimately prevailed over the worst global pandemic in a century. That does not, of course, guarantee that we will continue to be so fortunate indefinitely.

But whether or not the ultimate abolition of nuclear weapons is possible, as some advocate, there are concrete, practical steps that can be taken to reduce the dangers of accident, miscalculation, terrorism, and unrestrained proliferation that now face us. As detailed in the "A World Free of Nuclear Weapons" editorial, these include reducing nuclear arsenals from hair-trigger alert status, continuing to work for substantial arms reductions, increasing the security of existing weapons, and exerting firmer controls on nuclear fuel production.

Taking those steps, however, requires acknowledging the urgency of those threats. In a world in which the existence of nuclear weapons has become an accepted part of civilization, lost in the noise of other pressing concerns such as climate changes and social inequalities, the first challenge is increasing the level of public awareness and discussion.

Many people would undoubtedly and correctly cite climate change as a serious existential threat currently facing

There are concrete, practical steps that can be taken to reduce the dangers of accident, miscalculation, terror- ism, and unrestrained proliferation.

humanity. Yet nuclear weapons pose a threat just as global and far more immediate, not limited to any one nation, and one that transcends any political, social, or ideological concerns. They are the ultimate enforcer of equality, destroying with absolutely no regard to race, gender, political affiliations, or nationality. As such, they are everyone's problem. If we can't get rid of them, we must find a better way to live with them—one that relies on more than luck and fragile human nature.

There are two things, two major events or accomplishments or acts of humankind, for which the twentieth century will no doubt be remembered thousands of years hence, if human memory perseveres that long. One is venturing beyond the bounds of the earth to the moon. The other is the discovery and development of nuclear power, both in its benign civilian form for energy and darker incarnation as nuclear weapons. One may prove the path to our future; the other may be the harbinger of our extinction.

As we've seen, the idea of "humanity standing at the crossroads of extinction or survival" is one of the first rhetorical clichés of the atomic age. And yet it's still apt. As awesome and powerful as nuclear weapons may be, they are not some magical, supernatural force beyond human comprehension or control. We conceived nuclear weapons, decided to create them, and decided to continue building them.

If in the final analysis nuclear weapons are merely tools that we have invented and invested with certain real as well as unreal qualities, then we can also choose both how or whether to use them, and how to think about them. We don't need to accept the inevitability of doomsday. But to make those choices, it's necessary to acknowledge, accept, and consider all their implications and consequences. To paraphrase Abraham Lincoln, we can control nuclear weapons, or nuclear weapons will control us. The choice is ours.

antiballistic missile (ABM)
A weapon system designed to intercept and destroy attacking missiles.

arms control
Efforts to limit and control the production and deployment of nuclear weapons through the negotiation of treaty and verification measures.

Ballistic Missile Early Warning System (BMEWS)
A radar network to detect and track incoming Soviet missiles.

chain reaction
A self-sustaining fission process releasing large amounts of energy.

civil defense
Attempts to protect civilian populations through organized programs of public information, shelter construction, and disaster response.

Comprehensive Test Ban Treaty (CTBT)
Prohibits all nuclear weapons tests worldwide, including aboveground, underground, at sea, or in space. Adopted by the UN General Assembly in 1996, but not yet in force, and awaiting ratification by several signatories, including the United States.

counterforce
Targeting an enemy's military assets while excluding population centers and civilian areas.

countervalue
Targeting an enemy's cities and population centers to inflict maximum psychological and societal damage.

critical mass
An amount of fissionable material that will sustain a chain reaction under the proper conditions.

deterrence
The theory of possessing enough retaliatory power to dissuade attack from a potential enemy.

dirty bomb
A weapon that spreads radioactivity, but does not produce a nuclear explosion.

Distant Early Warning (DEW) Line
A string of radar stations, now deactivated, across the far northern borders of North America to detect Soviet attack.

doomsday machine
A hypothetical weapon that automatically destroys both sides if triggered by an attack or other specified circumstances.

electromagnetic pulse (EMP)
An extremely powerful burst of electromagnetic energy generated by a nuclear explosion that can temporarily or permanently disable electric and electronic equipment.

enrichment
Refining radioactive material such as uranium or plutonium to a state of purity that can sustain a chain reaction.

fail-safe
A system that automatically defaults to a "safe" mode, such as a weapon that will not fire or attack in case of accident or malfunction.

fallout
Radioactive atmospheric debris particles from a nuclear explosion.

fireball
A sphere of superheated gas and plasma formed at the onset of a nuclear explosion.

first strike
Initiating an attack against an enemy instead of responding only to retaliate.

fission

Splitting of the nucleus of a heavy element into smaller parts, with the release of great amounts of energy. Elements that do so readily are called fissionable or fissile materials.

flexible response

A strategy that incorporates multiple military options depending on various factors.

fusion

The combination of light nuclei into heavier elements, with the release of great amounts of energy.

gap

A perceived or actual imbalance in some type of military or technological capacity.

gun-type weapon

A nuclear weapon that rapidly forces together two or more pieces of subcritical mass to create a supercritical mass and resulting nuclear explosion.

implosion

The creation of a supercritical mass and nuclear explosion by the sudden compression of a mass of fissionable material.

intercontinental ballistic missile (ICBM)

A usually multistage guided missile that carries one or more nuclear warheads at intercontinental ranges of at least thirty-four hundred miles.

intermediate-range ballistic missile

A missile that carries nuclear warheads at a range of eighteen hundred to thirty-four hundred miles.

Intermediate-Range Nuclear Forces (INF) Treaty

A 1987 agreement, suspended in 2019, between the United States and USSR to abolish all land-based short- and intermediate-range missile systems.

International Atomic Energy Agency (IAEA)

International organization under UN auspices created in 1957 to promote peaceful atomic energy and control nuclear weapons.

isotope
A form of element with identical chemical properties, but a different atomic mass than another form of the same element, such as uranium-235 and uranium-238.

kiloton
Equivalent to one thousand tons of TNT; a measure of explosive force.

launch on warning
The policy of retaliation as soon as an attack is detected, without waiting for weapons to actually strike one's nation.

Limited Test Ban Treaty
Also known as the Partial Test Ban Treaty. Signed in 1963 by the United States, USSR, and United Kingdom. Bans nuclear testing in the atmosphere, at sea, and in space, limiting tests to underground only.

massive retaliation
A policy of responding to any nuclear attack or provocation, however limited, with full force.

megadeath
One million deaths; used as a shorthand term in nuclear war strategy and theory.

megaton
Equivalent to one million tons of TNT; a measure of explosive force.

multiple independently targetable reentry vehicle (MIRV)
Multiple nuclear warheads carried on a single missile that can each be guided to a separate target, allowing one ICBM to strike multiple targets.

mutual assured destruction (MAD)
A doctrine that any nuclear attack will inevitably result in the destruction of both sides.

preemptive/preventive war, strike
Attacking an enemy before they strike you, either on detecting an imminent attack (preemptive) or at a time of your choosing (preventive).

proliferation
The spread of nuclear weapons and technology to other nations.

reactor
A device to generate and control a nuclear chain reaction to create power or nuclear fuel material.

Strategic Air Command (SAC)
Strategic Air Command; part of the US Air Force formed in 1946, responsible for nuclear bombers and ICBMs. Disbanded in 1992.

Strategic Arms Limitation Talks (SALT)
Arms control negotiations and treaties (SALT I, SALT II, and ABM Treaty) between the United States and USSR. Succeeded by START and New START.

Strategic Arms Reduction Treaty (START)
Arms control agreements between the United States and USSR/Russia.

Strategic Defense Initiative (SDI)
1980s US system to defend against ICBMs. Also known as Star Wars.

strategic triad
Nuclear bombers, ICBMs, and submarine-launched missiles considered as a single military force, with three separate elements that ensure redundancy and survivability.

supercritical mass
The point at which a critical mass becomes enough to induce a nuclear explosion.

thermonuclear
A term for very high temperatures that induce nuclear fusion.

warhead
The explosive component of a missile.

yield
Total explosive energy of a nuclear weapon, usually expressed in kilotons/megatons, such as five kilotons, ten megatons, and so on.

NOTES

Chapter 1

1. "World Nuclear Weapon Stockpile," Ploughshares Fund, https://www.ploughshares.org/world-nuclear-stockpile-report.

2. See Xiaoping Yang, Robert North, Carl Romney, and Paul G. Richards, "Worldwide Nuclear Explosions," https://www.ldeo.columbia.edu/~richards/my_papers/WW_nuclear_tests_IASPEI_HB.pdf.

Chapter 2

1. Quoted in Richard Rhodes, *The Making of the Atomic Bomb* (New York: Simon and Schuster, 1986), 292.

2. Quoted in Cynthia C. Kelly, ed., *The Manhattan Project: The Birth of the Atomic Bomb in the Words of Its Creators, Eyewitnesses, and Historians* (New York: Black Dog and Leventhal, 2009), 43.

3. William L. Laurence, *Dawn over Zero: The Story of the Atomic Bomb* (New York: Alfred A. Knopf, 1946), 39.

4. Quoted in Rhodes, *The Making of the Atomic Bomb*, 305.

5. Quoted in Rhodes, *The Making of the Atomic Bomb*, 442.

6. Quoted in Rhodes, *The Making of the Atomic Bomb*, 442.

7. Some historians contend that physicist Werner Heisenberg, in charge of the German atomic program, deliberately stalled research in order to prevent Hitler from getting the atomic bomb. See, for example, Thomas Powers, *Heisenberg's War: The Secret History of the German Bomb* (New York: Da Capo Press, 2000).

8. Quoted in Kelly, *The Manhattan Project*, 339, 341.

Chapter 3

1. Quoted in Richard Rhodes, *Dark Sun: The Making of the Hydrogen Bomb* (New York: Simon and Schuster, 1995), 378.

2. Quoted in Daniel Ellsberg, *The Doomsday Machine: Confessions of a Nuclear War Planner* (New York: Bloomsbury, 2017), 290.

3. Quoted in Ellsberg, *The Doomsday Machine*, 291.

Chapter 4

1. Quoted in William I. Hitchcock, *The Age of Eisenhower: America and the World in the 1950s* (New York: Simon and Schuster, 2018), 208.

2. Quoted in Richard Rhodes, *Dark Sun: The Making of the Hydrogen Bomb* (New York: Simon and Schuster, 1995), 567.

3. Quoted in Karl E. Valois, ed., *The Cuban Missile Crisis: A World in Peril* (Carlisle, MA: Discovery Enterprises, 1998), 31.

4. Quoted in Stacey Bredhoff, *To the Brink: JFK and the Cuban Missile Crisis* (Washington, DC: Foundation for the National Archives, 2012), 54.

5. Also on Kennedy's mind was the fact that months earlier, he had ordered the removal from Turkey of the US missiles, which were obsolete and useless, but the delays in implementing his orders had left the United States susceptible to just this sort of pressure from the USSR, much to his annoyance.

6. Quoted in Scott D. Sagan, *The Limits of Safety: Organizations, Accidents, and Nuclear Weapons* (Princeton, NJ: Princeton University Press, 1993), 138.

7. *The Fog of War: Eleven Lessons from the Life of Robert S. McNamara*, directed by Errol Morris (Sony Pictures Classics, 2003).

8. Ironically, banishing testing to underground facilities only served to encourage more test shots while making them harder to detect, thus accelerating rather than slowing the arms race.

9. Quoted in Eric Schlosser, *Command and Control: Nuclear Weapons, the Damascus Accident, and the Illusion of Safety* (New York: Penguin, 2013), 352.

10. For descriptions of these incidents, see Sagan, *The Limits of Safety*.

Chapter 5

1. William L. Laurence, "Drama of the Atomic Bomb Found Climax in July 16 Test," *New York Times*, September 26, 1945, 1.

2. John Hersey, *Hiroshima* (New York: Vintage, 1989), 5–6, 8–9.

Chapter 6

1. The B-52 did almost that during its years over Southeast Asia, dropping well over seven million tons of bombs—an amount that, had it been unleashed all at once, would have been more than the equivalent of a thermonuclear strike. (Taking a cue from his former boss, Eisenhower, President Richard Nixon in fact hinted at using nuclear weapons in an attempt to bring North Vietnam to the negotiating table.)

2. "President Reagan's SDI Speech," March 23, 1983, Atomic Archive, https://www.atomicarchive.com/resources/documents/missile-defense/sdi-speech.html.

Chapter 7

1. Laurence's expertise in atomic matters not only made him more knowledgeable than his peers but also, in Groves's eyes, perhaps more dangerous.

Bringing the reporter under the aegis of the top secret world of the project was a convenient way of making certain that he didn't say or reveal the wrong thing in the pages of the *New York Times*.

2. See Fred Kaplan, *The Bomb: Presidents, Generals, and the Secret History of Nuclear War* (New York: Simon and Schuster, 2020), 166.

3. The film revolves around what was being referred to at that time as the ideal "nuclear" family, consisting of a husband, wife, and two teenage children—one male, and one female.

4. See Dawn Stover, "Facing Nuclear Reality, 35 Years after *The Day After*," *Bulletin of the Atomic Scientists*, https://thebulletin.org/facing-nuclear-reality-35-years-after-the-day-after/.

Chapter 8

1. Owen B. Toon, Charles G. Bardeen, Alan Robock, Lili Xia, Hans Kristensen, Matthew McKinzie, R. J. Peterson, et al., "Rapidly Expanding Nuclear Arsenals in Pakistan and India Portend Regional and Global Catastrophe," *Science Advances* 5, no. 10 (October 2, 2019), doi: 10.1126/sciadv.aay5478.

2. See, for example, John Lewis Gaddis, "The Long Peace: Elements of Stability in the Postwar International System," *International Security* 10, no. 4 (Spring 1986): 99–142.

3. Stephen M. Younger, *The Bomb: A New History* (New York: HarperCollins, 2009), 117, 122.

4. Ward Wilson, *Five Myths about Nuclear Weapons* (Boston: Houghton Mifflin Harcourt, 2013), 116, 121.

5. "JFK American University Commencement Address," June 10, 1963, Washington, DC, American Rhetoric, https://www.americanrhetoric.com/speeches/jfkamericanuniversityaddress.html.

6. Quoted on the DVD *Nuclear Tipping Point*, Nuclear Security Project, 2010, https://www.nti.org/about/projects/nuclear-tipping-point/.

7. George P. Shultz, William J. Perry, Henry A. Kissinger, and Sam Nunn, "A World Free of Nuclear Weapons," *Wall Street Journal*, January 4, 2007, A15.

8. Shultz et al., "A World Free of Nuclear Weapons."

9. Quoted in Mark Wolverton, "Scientists and Strategists Contemplate the Increasing Odds of Nuclear War," *Undark*, April 18, 2017, https://undark.org/2017/04/18/contemplating-nuclear-armageddon-war/.

FURTHER READING

This is only a partial list out of many recommended resources that have gone into the making of this book, but it provides a representative sample of the vast material available on this subject. Note also that this classification is somewhat arbitrary; many of these works fit into several categories.

Technicalities

Baker, David. *Nuclear Weapons: 1945 Onwards (Strategic and Tactical Delivery Systems)*. Somerset, UK: Haynes Publishing, 2017.

Bernstein, Jeremy. *Nuclear Weapons*. New York: Cambridge University Press, 2010.

Glasstone, Samuel, and Philip J. Dolan. *The Effects of Nuclear Weapons*. Washington, DC: US Government Printing Office, 1977.

Serber, Robert. *The Los Alamos Primer*. Berkeley: University of California Press, 1992.

Smyth, Henry DeWolf. *Atomic Energy for Military Purposes: The Official Report on the Development of the Atomic Bomb under the Auspices of the United States Government 1940–1945*. Princeton, NJ: Princeton University Press, 1945.

History

Ambinder, Marc. *The Brink: President Reagan and the Nuclear War Scare of 1983*. New York: Simon and Schuster, 2018.

Badash, Lawrence. *Scientists and the Development of Nuclear Weapons: From Fission to the Limited Test Ban Treaty 1939–1963*. Amherst, NY: Prometheus Books, 1995.

Bird, Kai, and Martin J. Sherwin. *American Prometheus: The Triumph and Tragedy of J. Robert Oppenheimer*. New York: Vintage Books, 2006.

Blight, James G., and Janet M. Lang. *The Fog of War: Lessons from the Life of Robert S. McNamara*. Lanham, MD: Rowman and Littlefield, 2005.

Boyer, Paul. *By the Bomb's Early Light: American Thought and Culture at the Dawn of the Atomic Age*. New York: Pantheon, 1985.

Brands, H. W. *The General and the President: MacArthur and Truman at the Brink of Nuclear War*. New York: Anchor, 2016.

Bredhoff, Stacey. *To the Brink: JFK and the Cuban Missile Crisis*. Washington, DC: Foundation for the National Archives, 2012.

Bundy, McGeorge. *Danger and Survival: Choices about the Bomb in the First Fifty Years*. New York: Random House, 1988.

Hersey, John. *Hiroshima*. New York: Vintage, 1989.

Hewlett, Richard G., and Oscar E. Anderson Jr. *The New World, 1939/1946: A History of the United States Atomic Energy Commission, Volume I*. University Park: Pennsylvania State University Press, 1962.

Hewlett, Richard G., and Francis Duncan. *Atomic Shield, 1947/1952: A History of the United States Atomic Energy Commission, Volume II*. University Park: Pennsylvania State University Press, 1969.

Hewlett, Richard G., and Jack M. Holl. *Atoms for Peace and War, 1953/1961: A History of the United States Atomic Energy Commission, Volume III*. Berkeley: University of California Press, 1989.

Hitchcock, William I. *The Age of Eisenhower: America and the World in the 1950s*. New York: Simon and Schuster, 2018.

Hoffman, David E. *The Dead Hand: The Untold Story of the Cold War Arms Race and Its Dangerous Legacy*. New York: Doubleday, 2009.

Jones, Nate, ed. *Able Archer 83: The Secret History of the NATO Exercise That Almost Triggered Nuclear War*. New York: New Press, 2016.

Kaplan, Fred. *The Bomb: Presidents, Generals, and the Secret History of Nuclear War*. New York: Simon and Schuster, 2020.

Kaplan, Fred. *The Wizards of Armageddon: The Untold Story of the Small Group of Men Who Have Devised the Plans and Shaped the Policies on How to Use the Bomb*. Stanford, CA: Stanford University Press, 1991.

Keeney, L. Douglas. *The Doomsday Scenario: The Official Doomsday Scenario Written by the United States Government during the Cold War*. Saint Paul, MN: MBI Publishing, 2002.

Keeney, L. Douglas. *15 Minutes: General Curtis LeMay and the Countdown to Nuclear Annihilation*. New York: St. Martin's Press, 2011.

Kelly, Cynthia C., ed. *The Manhattan Project: The Birth of the Atomic Bomb in the Words of Its Creators, Eyewitnesses, and Historians*. New York: Black Dog and Leventhal, 2009.

Kennedy, Robert F. *Thirteen Days: A Memoir of the Cuban Missile Crisis*. New York: W. W. Norton, 1971.

Lamont, Lansing. *Day of Trinity*. New York: Scribner, 1965.

Laurence, William L. *Dawn over Zero: The Story of the Atomic Bomb*. New York: Alfred A. Knopf, 1946.

Laurence, William L. *Men and Atoms: The Discovery, the Uses and the Future of Atomic Energy*. New York: Simon and Schuster, 1959.

Miller, Richard L. *Under the Cloud: The Decades of Nuclear Testing*. Woodlands, TX: Two-Sixty Press, 1991.

Perry, William J. *My Journey at the Nuclear Brink*. Stanford, CA: Stanford University Press, 2015.

Rhodes, Richard. *Dark Sun: The Making of the Hydrogen Bomb*. New York: Simon and Schuster, 1995.

Rhodes, Richard. *The Making of the Atomic Bomb*. New York: Simon and Schuster, 1986.

Rosenbaum, Ron. *How the End Begins: The Road to a Nuclear World War III*. New York: Simon and Schuster, 2011.

Sherwin, Martin J. *Gambling with Armageddon: Nuclear Roulette from Hiroshima to the Cuban Missile Crisis*. New York: Alfred A. Knopf, 2020.

Sherwin, Martin J. *A World Destroyed: Hiroshima and Its Legacies*. Stanford, CA: Stanford University Press, 2003.

Siracusa, Joseph M. *Nuclear Weapons: A Very Short Introduction*. Oxford: Oxford University Press, 2015.

Smith, P. D. *Doomsday Men: The Real Dr. Strangelove and the Dream of the Super-weapon*. New York: St. Martin's Press, 2007.

Stern, Sheldon M. *The Cuban Missile Crisis in American Memory: Myths versus Reality*. Stanford, CA: Stanford University Press, 2012.

Taubman, Philip. *The Partnership: Five Cold Warriors and Their Quest to Ban the Bomb*. New York: HarperCollins, 2012.

Thomas, Evan. *Ike's Bluff: President Eisenhower's Secret Battle to Save the World*. New York: Little, Brown, 2012.

Tucker, Todd. *Atomic America: How a Deadly Explosion and a Feared Admiral Changed the Course of Nuclear History*. New York: Free Press, 2009.

Valois, Karl E., ed. *The Cuban Missile Crisis: A World in Peril*. Carlisle, MA: Discovery Enterprises, 1998.

Wolverton, Mark. *Burning the Sky: Operation Argus and the Untold Story of the Cold War Nuclear Tests in Outer Space*. New York: Overlook Press, 2018.

Wolverton, Mark. *A Life in Twilight: The Final Years of J. Robert Oppenheimer*. New York: St. Martin's Press, 2008.

Wyden, Peter. *Day One: Before Hiroshima and After*. New York: Simon and Schuster, 1984.

York, Herbert F. *Making Weapons, Talking Peace: A Physicist's Odyssey from Hiroshima to Geneva*. New York: Basic Books, 1987.

Younger, Stephen M. *The Bomb: A New History*. New York: HarperCollins, 2009.

Popular Culture

The Atomic Cafe. Directed by Kevin Rafferty, Jayne Loader, Pierce Rafferty. Archives Project, 1982.

Barson, Michael. *Better Dead Than Red! A Nostalgic Look at the Golden Years of Russiaphobia, Red-Baiting, and Other Commie Madness*. New York: Hyperion, 1992.

Barson, Michael, and Steven Heller. *Red Scared! The Commie Menace in Propaganda and Popular Culture*. San Francisco: Chronicle Books, 2001.

The Beginning or the End. Directed by Norman Taurog. Warner Brothers, 1947.

Bryant, Peter. *Red Alert*. New York: Ace, 1958.

Burdick, Eugene, and Harvey Wheeler. *Fail-Safe*. New York: Dell, 1962.

Colossus: The Forbin Project. Directed by Joseph Sargent. Universal, 1970.

The Day After. Directed by Nicholas Meyer. ABC Circle Films, 1983.

The Day After Trinity: J. Robert Oppenheimer and the Atomic Bomb. Directed by Jon Else. KTEH Public Television, 1981.

The Day the Earth Stood Still. Directed by Robert Wise. 20th Century Fox, 1951.

Dr. Strangelove, or: How I Learned to Stop Worrying and Love the Bomb. Directed by Stanley Kubrick. Columbia Pictures, 1964.

Fail-Safe. Directed by Sidney Lumet. Columbia Pictures, 1964.

The Fog of War: Eleven Lessons from the Life of Robert S. McNamara. Directed by Errol Morris. Sony Pictures Classics, 2003.

Gojira. Directed by Ishiro Honda. Toho Studios, 1954.

Newkey-Burden, Chas. *Nuclear Paranoia*. Harpenden, UK: Pocket Essentials, 2003.

On the Beach. Directed by Stanley Kramer. United Artists, 1959.

Panic in Year Zero! Directed by Ray Milland. American International Pictures, 1962.

Swedin, Eric G. *Survive the Bomb: The Radioactive Citizen's Guide to Nuclear Survival*. Minneapolis: Zenith Press, 2011.

Them! Directed by Gordon I. Douglas. Warner Brothers, 1951.

Threads. Directed by Mick Jackson. BBC, 1984.

The Trials of J. Robert Oppenheimer. Directed by David Grubin. WGBH Boston, 2009.

Trinity and Beyond: The Atomic Bomb Movie. Directed by Peter Kuran. Visual Concept Entertainment, 2000.

The War Game. Directed by Peter Watkins. BBC, 1966.

Weart, Spencer. *Nuclear Fear: A History of Images*. Cambridge, MA: Harvard University Press, 1988.

Wells, H. G. *The World Set Free: A Story of Mankind*. New York: E. P. Dutton, 1914.

Policy, Strategy, and Safety

Allison, Graham. *Nuclear Terrorism: The Ultimate Preventable Catastrophe*. New York: Times Books, 2004.

Ellsberg, Daniel. *The Doomsday Machine: Confessions of a Nuclear War Planner*. New York: Bloomsbury, 2017.

Perry, William J., and Tom Z. Collina. *The Button: The New Nuclear Arms Race and Presidential Power from Truman to Trump*. Dallas: BenBella Books, 2020.

Rubinson, Paul. *Redefining Science: Scientists, the National Security State, and Nuclear Weapons in Cold War America*. Amherst: University of Massachusetts Press, 2016.

Sagan, Scott D. *The Limits of Safety: Organizations, Accidents, and Nuclear Weapons*. Princeton, NJ: Princeton University Press, 1993.

Sagan, Scott D., and Kenneth N. Waltz. *The Spread of Nuclear Weapons: A Debate Renewed*. New York: W. W. Norton, 2003.

Schelling, Thomas. *Arms and Influence*. New Haven, CT: Yale University Press, 2008.

Schlosser, Eric. *Command and Control: Nuclear Weapons, the Damascus Accident, and the Illusion of Safety*. New York: Penguin, 2013.

Wilson, Ward. *Five Myths about Nuclear Weapons*. Boston: Houghton Mifflin Harcourt, 2013.

INDEX

The MIT Press Essential Knowledge Series

MARK WOLVERTON has written widely on the history of the Cold War and nuclear weapons. He is the author of *Burning the Sky: Operation Argus and the Untold Story of the Cold War Nuclear Tests in Outer Space* and *A Life in Twilight: The Final Years of J. Robert Oppenheimer*, among several other books.